THE RATIONAL
AND THE SOCIAL

PHILOSOPHICAL ISSUES IN SCIENCE

Edited by W. H. Newton-Smith
Balliol College, Oxford

THE RATIONAL
AND THE SOCIAL

JAMES ROBERT BROWN

Associate Professor in the Department of Philosophy,
University of Toronto, Canada

R

ROUTLEDGE
London and New York

For those I grew up with

my father, Andrew
my mother, Isabel
my brother, Alan
my sister, Andrée
my brother, Gordon

First published in 1989 by Routledge
11 New Fetter Lane, London EC4P 4EE
29 West 35th Street, New York NY 10001

© 1989 James Robert Brown

Typeset in Baskerville by Columns of Reading

Printed in Great Britain by
T. J. Press (Padstow) Ltd, Padstow, Cornwall

British Library Cataloguing in Publication Data

Brown, James Robert
The rational and the social. ——
(Philosophical issues in science).
1. Philosophy of science. Theories
I. Title   II. Series
501

Library of Congress Cataloging in Publication Data

Brown, James Robert
The rational and the social. ——
(Philosophical issues in science).
Bibliography: p.
Includes index.
1. Knowledge, Sociology of. 2. Science—Philosphy.
3. Science—Social aspects. 4. Rationalism.
5. History—Philosophy. I. Title. II. Series.
BD175.B76   1989   121   88–24211

ISBN 0–415–02905–8

CONTENTS

PREFACE

Ever since Plato took up the cudgels against Protagoras, when the latter said that 'man is the measure of all things', the battle over relativism has been fought again and again. It may seem like an endless cycle of claims and counter-claims, but we would be mistaken to think that nothing is ever learned in any of the successive clashes.

Among the main contenders in the present debate are, on the one hand, a small but vigorous and influential group of sociologists and historians of science centred in Edinburgh. They preach (and practise) a radically sociological approach to the understanding of how knowledge (scientific knowledge, in particular) is acquired. On the other side of this debate are their rationalist opponents, including most philosophers of science and traditional historians of ideas. They see 'evidence', 'good reasons', and 'rational belief' rather than non-cognitive 'interests' as the guiding force behind the development of science.

This book is a contribution to the present debate. It is, for the most part, squarely on the rationalist side. The first chapter is an exposition of the new sociological way of doing things, the so-called 'strong programme for the sociology of knowledge'. This chapter is followed by two more devoted to critically examining some of the chief tenets of the sociological movement. David Bloor's 'science of science' and Barry Barnes's 'finitism' are the main targets of chapters two and three, respectively. Some special topics in experimentation come up in the fourth chapter. The anthropological metaphor comes into play in the fifth and especially the sixth chapters, which are mainly about the role history might play in determining just what rationality is. The account given leads to a richer, and I think proper, way of understanding how science works. It is at this point that we begin to get a glimmer of how the social really enters science. Its penetration is quite significant, but its *locus* is not where the present debate between rationalists and sociologists is usually focused. The final chapter is devoted to this issue and, in particular, to what we might do about it; that is, it's about philosophical interventions to make science better.

That, in brief, is what the present book is about, but a preface should say something more; it should also say something about the author's larger intentions and motivations. I find that I have several. One of my concerns is to defend the old ideal of 'knowledge for its own sake'; but holding forth on this invariably induces cynicism or yawning, so the topic is best avoided. (Of course I do not wish to defend a childish view of the purity of science, and would readily admit that sociologists have taught us much about the sorts of things we must avoid in defending such an ideal.)

The other motivation has to do with political concerns. I am convinced that the social goals many of us hanker after, namely, a socialist world with equality among the classes, sexes, and races, is best served by rationality. It is true that some scientific theories (often pseudo-scientific, I would claim) have done considerable harm. Lots of biological theories, for instance, have been racist or sexist or have been used in defence of the most pernicious forms of free enterprise. This is undoubtedly why so many who share my political goals find themselves attracted to the sociologists' account of science. (There are speculations along this line about the sociologists' motives in the second chapter.) I am quite persuaded, however, that such a view is mistaken, even dangerously so, for the scepticism and relativism which result from it tend to produce a quietism and inaction rooted in a sense of hopelessness and pointlessness. One of the great forces for evil today is the religious right. Its members clamour for equal time with Darwin for 'special creation', since after all, both are 'just theories', as Ronald Reagan put it. This, of course, is merely a front for a not-so-hidden agenda; the suppression of women, trade unions, and third-world people is part of the bargain. Several relativist-minded sociologists of knowledge despise these political trends just as I do, but how are they to fight the battle? By saying, 'Yes, all theories are equally good; they merely serve different social interests'? Surely we can do better than that.

On the other hand, science should not be left to its own devices. It can be improved, and I shall try to say how in the final chapter. But in spite of its shortcomings, science, on balance, is a force for liberation (as well as being a joy to behold in its own right). Perhaps this book will go a little way toward convincing some readers of this fact. We have every good reason to think that rationality serves the demands of justice just as it serves the demands of curiosity. Corny as it may sound, science is a friend of the oppressed just as surely as it is the glorious entertainer.

Well, enough of the missionary talking to cannibals; let's get on with it.

ACKNOWLEDGEMENTS

Writing a book makes one aware just how much is owed to others. Morsels of information and good sense obtained by osmosis often make their way to the front of the mind when the time requires it. How much I owe to others I cannot say, but I do not doubt that there are traces on every page. But, inevitably, only some of these can be singled out for explicit thanks: Ian Hacking, Noretta Koertge, Larry Laudan, Andrew Lugg, John Nicholas, Alex Rosenberg, and Michael Ruse. I also learned a great deal from the contributors to two recent volumes, Hollis and Lukes (1982) and Brown (1984). There are others to whom I owe even more. They include: William Newton-Smith, who invited the book in the first place, offered several valuable suggestions for improvement, and displayed great patience at my slow progress; David Bloor, who has been a most valuable correspondent of long standing and who spent a memorable day riding trains and sitting in Paddington station arguing these issues with me; and Robert Butts, with whom I have had numerous valuable conversations on these issues over the last dozen years. My greatest debt of all is to Kathleen Okruhlik, with whom I have argued these issues for longer than either of us cares to remember. Several ideas have been borrowed from her, only a few with acknowledgement. She read the penultimate draft and greatly improved it. To these and to those I haven't named, many thanks.

Chapter One

THE SOCIOLOGICAL TURN

The problem we are concerned with is just this: How should we understand science? Are we to account for scientific knowledge (or rather, belief)[1] by appeal to the various social factors which may have been prevalent when the theory was being formulated? That is, should we appeal to the 'interests' which a group of scientists may have had? Undoubtedly, social factors play some role, but are social causes totally responsible for the production of belief? Or should we instead take a different approach and account for scientific knowledge in a fashion which mirrors the very accounts that scientists themselves might have typically given to justify their theory choices? Perhaps we should be citing the 'evidence' for the beliefs in question; perhaps we should be providing 'good reasons' as part of the explanation for holding the belief. Which approach to understanding science is right?

The most effective modern champions of social causation are the members of the 'Edinburgh School', a very vigorous group of sociologists and historians of science largely centred in the Science Studies Unit of the University of Edinburgh. The principal opposition to this way of viewing science comes from philosophers as well as from the more traditional sociologists and intellectual historians. But the problem we are faced with is as old as it is tricky; so before elaborating on the modern debate, a backward glance would not be out of place.

THE NATURAL VERSUS THE SOCIAL SCIENCES

If we are to set the stage properly for the concerns of this book, we could hardly do better than to start with a brief look at Karl

Mannheim's view. In *Ideology and Utopia*, the *magnum opus* of the father of the modern sociology of knowledge, Mannheim characterizes the discipline this way: 'The principal thesis of the sociology of knowledge is that there are modes of thought which cannot be adequately understood as long as their social origins are obscured' (Mannheim 1936:2). In itself, this seems quite innocuous; it expresses a sentiment that we all might agree to, for we all concur that *some* people believe *some* of the things they do because of social factors. Quarrels do not arise until it becomes a question of *which* people and *which* beliefs. Things usually become heated when it is suggested that all of our *scientific* beliefs have a (possibly contaminating) social origin.

There has been a long tradition in the sociology of knowledge which has sharply separated beliefs into two kinds. A boundary is drawn between mathematics and the natural sciences on the one hand, and almost everything else on the other. Included in this latter collection are such 'cultural' things as: religious beliefs, morals, 'practical wisdom', and, often enough, the social sciences. The natural sciences are viewed as pristine, uncorrupted by any considerations of interest, while everything else is at least suspect, viewed as likely ideology, tarnished with subjectivity, and corrupted with naked or concealed self-interest.

Mannheim is part of this dualistic tradition (at least in some of his writings; he tended to be ambivalent on the issue). In one place he expresses the dichotomy this way:

> It may be said for formal knowledge that it is essentially accessible to all and that its content is unaffected by the individual subject and his historical-social affiliations. But, on the other hand, it is certain that there is a wide range of subject-matter which is accessible only either to certain subjects, or in certain historical periods, and which becomes apparent through the social purposes of individuals.
>
> (Mannheim 1936: 150)

By 'formal knowledge' he means the natural sciences and mathematics. In another passage, Mannheim again gives voice to the duality between the natural and the social:

> Are the existential factors in the social process merely of peripheral significance, are they to be regarded merely as

2

conditioning the origin or factual development of ideas (i.e. are they of merely genetic relevance), or do they penetrate into the 'perspective' of concrete particular assertions? This is the next question we shall try to answer. The historical and social genesis of an idea would only be irrelevant to its ultimate validity if the temporal and social conditions of its emergence had no effect on its content and form. If this were the case, any two periods in the history of human knowledge would only be distinguished from one another by the fact that in the earlier period certain things were still unknown and certain errors still existed which, through later knowledge were completely corrected. This simple relationship between an earlier incomplete and a later complete period of knowledge may to a large extent be appropriate for the exact sciences (although indeed today the notion of the stability of the categorical structure of the exact science is, compared with the logic of classical physics, considerably shaken). For the history of the cultural sciences, however, the earlier stages are not quite so simply superseded by the later stages, and it is not so easily demonstrable that early errors have subsequently been corrected. Every epoch has its fundamentally new approach and its characteristic point of view, and consequently sees the 'same' object from a new perspective.

(Mannheim 1936: 243)

Admittedly, Mannheim hedges on the 'stability of the exact sciences'; nevertheless, a sharp distinction between such things as physics, chemistry, and mathematics on the one hand, and what he calls the 'cultural sciences' on the other, is made.

In consequence of this distinction, we should have a similar dichotomy in our approach to accounting for belief. If we wish to explain why a certain scientist or community of scientists believes (or did in the past believe) a particular theory of physics (or some other natural science) we should look to the evidential reasons which were available. However, if it is a belief in a theological, moral, or economic doctrine that is to be accounted for, then 'evidence' would have little or nothing to do with it. Instead, we must look to the social factors prevalent at the time the theory choice was made, for it is these social factors which would have caused such a decision. We should point to the evidence to account for the growth of natural science; and we should point to social causes to account for every other kind of belief.

3

It is very important to notice, in the passage cited, why Mannheim makes the distinction between the cultural and the natural sciences. He thinks that the exact or natural sciences exemplify a 'simple relationship between an earlier incomplete and a later complete period of knowledge'. This reflects a view of the development of science known as the 'cumulative' account: Once a fact has been discovered, or a theory established, it is never abandoned; only more facts are added to it.

Mannheim is not alone in holding such a view of natural science; nor is he alone in thinking it is a good reason for a hands-off attitude toward it by sociologists. In a recent exegetical piece on the sociology of knowledge, Werner Stark voices the same sentiment when he writes, 'Because man must take the facts of nature as he finds them, while the facts of culture are his own work, the social determination of knowledge will be different in the two instances' (Stark 1967: 477, vol. 7).

This is, however, a view of the natural sciences that few today give any credence to. In fact, quite the contrary. Whether they are proponents of a 'rational' or of a 'sociological' approach to the understanding of science, virtually all contemporary commentators think the natural sciences have been revolutionary, not cumulative. Many past theories and even many past 'facts' have been completely overthrown.

Indeed, attacks (especially that found in Kuhn's *The Structure of Scientific Revolutions*) on the cumulative account of science have done much to inspire the contemporary sociological turn. One of the leading figures in the recent sociological movement, Barry Barnes, puts it this way:

It is well known that as scientific knowledge has developed, numerous mechanisms and theories have been postulated and successively set aside. This is, indeed, why so many philosophers of science have struggled to maintain a fact/theory distinction, and to base their justificatory rhetoric on the accumulation of facts. But there has also been a good deal of informal faith placed in the progressive quality of this sequence of theories and mechanisms. Recent historical studies, however, in particular those of T. S. Kuhn (1970), effectively undermine this faith; they demonstrate that fundamental theoretical transitions in science are not simply rational responses to increased knowledge of

4

reality, predictable in terms of context-independent standards of inference and evaluation.

(Barnes 1977: 23)

Most philosophers would not put it as Barnes has done, but they would agree with the upshot: Mannheim's reason for distinguishing between the natural sciences and the social sciences is not such a good reason after all. The natural sciences do not develop in a cumulative fashion. But does this then mean that we must look to social causes to explain the developments in the natural as well as the social sciences? Must we explain *all* beliefs by citing social factors? Or should we never turn to the social to account for any beliefs at all and look instead for 'good reasons' to account for every sort of belief? Perhaps there are yet other ways to look at it.

THE ARATIONALITY PRINCIPLE

Though the dichotomy between the natural and the social sciences is not one which will stand up, nevertheless, its breakdown need not lead automatically to a fully-fledged sociological account of *all* belief. One might maintain that what Kuhn and other historians have shown is not that the natural sciences require a sociological account, but rather that we need a new and richer notion of rationality. It will have to be a notion which, among other things, does not require a cumulative history of science. And this is just what several contemporary, post-positivistic, philosophers of science have been trying to do.

Larry Laudan is one prominent philosopher who is working in this direction. His *Progress and its Problems* contains a number of interesting suggestions and proposals, but one of the most important is his denial of any sharp boundary between science and non-science, or between the natural and the social sciences. What is good method in science, he claims, is good method anywhere that there are cognitive aims. Thus, physics and theology, mathematics and metaphysics, geology and economics are all on a par as far as they have the same method for correctly pursuing their cognitive ends. Of course, the practitioners of one or the other of these disciplines may not be following the method properly, but if the various practitioners were to be completely rational they would use the same general procedures. Scientific rationality simply means following the right method, and it is the same method for all.

As well as rejecting a demarcation between science and other cognitive activities, Laudan would also reject the idea that the practitioners of the natural sciences have made nothing but rational choices. Indeed, no one seriously believes that every scientist who ever held to some theory of physics, chemistry, or mathematics, did so because of the evidence then available. It is widely thought, even by the staunchest champions of scientific rationality, that sometimes a scientist will act irrationally, either by believing something which is totally crazy or by believing the right thing for the wrong reasons. And those 'bad reasons' might often stem from social forces. It will simply not be true that every cognitive decision in the history of the natural sciences can be accounted for by appeal to the evidence available to those who made the decision.

In consequence, some sort of guideline is required, some sort of rule which will tell the historian how to approach individual cases in the history of science. A rule which has been proposed is this: If a belief can be explained as being the result of the rational examination of the evidence available then that should be accepted as the correct explanation. If, and only if, no such rational explanation is available, should we account for the belief by appeal to a social (or some other) cause. In the frequently used jargon of 'internal/external', we should only seek an external account if no internal one can be found. Robert Merton, for instance, holds such a view: 'thought has an existential [i.e., social] basis in so far as it is not immanently [i.e., rationally] determined' (1969:516). This methodological principle is explicitly adopted by Laudan and called the 'arationality principle':

> basically, it amounts to the claim that *the sociology of knowledge may step in to explain beliefs if and only if those beliefs cannot be explained in terms of their rational merits*. . . . Essentially, the arationality assumption establishes a division of labor between the historian of ideas and the sociologist of knowledge; saying, in effect, that the historian of ideas, using the machinery available to him, can explain the history of thought insofar as it is rationally well-founded and that the sociologist of knowledge steps in at precisely those points where a rational analysis of the

6

acceptance (or rejection) of an idea fails to square with the
actual situation.

(Laudan 1977: 202)

If a slogan is useful, let it be this: 'Sociology is only for deviants'
(Newton-Smith 1981:238). The sociologist is to step in when
and only when there is some deviation from the norm of
rationality.

The arationality principle is probably the very antithesis of the
sentiments embodied in the recent sociological turn. Though I do
not accept it myself without major qualifications, its advocacy and
employment are common. As a rule of thumb, it embodies the
rationalist outlook. It is interesting that the (traditional) sociologist
Robert Merton and the rationalist philosopher Larry Laudan
concur in their endorsement of this approach to understanding
science. The recent sociological turn is as much an attack on the
Mertonian way of doing the sociology of science as it is an attack
on philosophers of science and traditional historians of ideas. The
lines of battle and the disciplinary boundaries do not correspond.

THE STRONG PROGRAMME

The claims of the new cognitive sociologists of knowledge,
especially the members of the Edinburgh school, are much stronger
than the mere assertion that sometimes social factors have to be
considered in order to have a complete account of an episode in the
history of science. Rather, it is maintained that social causes are
always present; they are the determining factors. This position is
most clearly and forcefully put in David Bloor's important and
influential work *Knowledge and Social Imagery*, and it goes by the
name 'The Strong Programme'.

Broadly speaking, there seem to be three types of consideration
used to support the new sociological approach. The first of these is
based on the claim that it is the only approach to science which is
itself scientific. The second stems from philosophical considerations
about underdetermination and related issues, the suggestion being
that there is not enough evidence to make rational decisions
anyway. The third type of alleged support for the sociological
approach comes from the perceived successes of recent case studies.

I will briefly describe each of these three considerations now, and take them up in detail critically in the following chapters.

Bloor chastises many of his fellow (traditional) sociologists for 'a betrayal of their disciplinary standpoint' (Bloor 1976:1). The stance of many traditional sociologists who take a hands-off attitude toward science is anathema to him. They are unnecessarily limiting the scope of their own enterprise; they should bring their considerable resources to bear in the very content of scientific knowledge. Philosophers, according to Bloor, have traditionally given sociologists only the non-natural sciences to account for, or only the irrational residue to explain, or when philosophers distinguish between 'discovery' and 'justification', the sociologists are given only the former to deal with. But, asserts Bloor, *all* of science is in the legitimate domain of the sociologist. He thus proposes the following tenets as characterizing the right way to do the sociology of knowledge and the only way to properly understand science (Bloor 1976: 5):

1. *Causality*. A proper account of science would be causal, that is, concerned with the conditions that bring about belief or states of knowledge.
2. *Impartiality*. It would be impartial with respect to truth and falsity, rationality or irrationality, success or failure. Both sides of these dichotomies will require explanation.
3. *Symmetry*. It would be symmetrical in its style of explanation. The same types of cause would explain, say, true and false, [rational and irrational, successful and unsuccessful] beliefs.
4. *Reflexivity*. It would be reflexive. In principle its patterns of explanation would have to be applicable to sociology itself. Like the requirement of symmetry, this is a response to the need to seek for general explanations. It is an obvious requirement of principle, otherwise sociology would be a standing refutation of its own theories.

Bloor has much to say in defence of each of these tenets of the 'strong programme'. The main point is that the sociologist is a scientist too, and ought to act as scientists do; he or she should try to characterize knowledge in a scientific fashion. 'If sociology could not be applied in a thorough-going way to scientific knowlege it would mean', concludes Bloor, 'that science could not scientifically know itself' (Bloor 1976:40). Thus, he says of the sociologist:

His ideas therefore will be in the same causal idiom as any other scientist. His concern will be to locate the regularities and general principles or processes which appear to be at work within the field of his data. His aim will be to build theories to explain these regularities. If these theories are to satisfy the requirement of maximal generality they will have to apply to both true and false beliefs, and as far as possible the same type of explanation will have to apply in both cases. The aim of physiology is to explain the organism in health and disease; the aim of mechanics is to understand machines which work and machines which fail; bridges which stand as well as those which fall. Similarly the sociologist seeks theories which explain the beliefs which are in fact found, regardless of how the investigator evaluates them.

(Bloor 1976:3)

In other words, to be truly scientific, which Bloor certainly takes to be a good thing, one has to look for the causes of beliefs. Moreover, we do not have two theories of nature, one for explaining why a bridge stands up (when it does) and a second theory for explaining why another bridge has fallen down. We have but one theory and we use it impartially and symmetrically to explain both standing and fallen bridges.

Barry Barnes makes methodological claims similar to the symmetry principle when he insists that:

What matters is that we recognize the *sociological* equivalence of different knowledge claims. We will doubtless continue to evaluate beliefs differently ourselves, but such evaluations must be recognized as having no relevance to the task of sociological explanation; as a methodological principle we must not allow our evaluation of beliefs to determine which form of sociological account we put forward to explain them.

(Barnes 1977:25)

Strong programmers, such as Bloor and Barnes, are not alone in making these kinds of demands on any acceptable account of science. Recent feminist critics are voicing similar complaints against traditional analyses of scientific knowledge. Sandra Harding protests that science 'excludes itself from the categories and activities it prescribes for everything else. It recommends that we understand everything but science through causal analyses and critical scrutiny of inherited beliefs' (Harding 1986:36). She further remarks that a '*thoroughgoing* and *scientific* appreciation of science requires descriptions and explanations of the regularities and underlying causal tendencies of science's own social practices and beliefs'. Harding then asks rhetorically, 'To what other "community of natives" would we give the final word about the causes, consequences, and social meanings of their own beliefs and institutions?' (1986:39).

A common argument directed against such a full-blooded sociology of scientific knowledge goes like this: If all beliefs are caused by social factors then this must be true of the strong programme as well. Therefore, the strong programme falls into a self-refuting relativism; it undermines its own position. Bloor, however, thinks this is not damaging in the least. The principle of reflexivity is simply an admission of the premiss of this argument. But the conclusion does not follow, says Bloor, unless social determination implies falsehood. But causation does not imply error, so the charge of 'self-refutation' will not stick.

There is one more consideration, stressed by Steven Shapin. This is the demand that to be scientific is to take correlations seriously. When a correlation is found, for instance, between a belief and a social class then it must be explained. Often, says Shapin, the explanation will be in terms of social factors. The 'scientific' character of the recent sociological turn will be the subject of the next chapter. But the requirement of being 'scientific' is not the only motivation for the recent sociological turn.

Another factor which has played a big role stems from recent (largely philosophical)[2] work on meaning and reference and the problems of 'underdetermination' and 'incommensurability'. The problem of underdetermination, briefly, is this: there are indefinitely many logically possible theories which are compatible with the given empirical data. Thus, the experimental evidence cannot pick out one from among these as being the uniquely correct or true theory. So, it is sometimes concluded, the decision which scientists actually do make cannot be based on rational considerations. Therefore, we must look elsewhere for the causes of the choice. As Bloor puts it: 'the theoretical component of knowledge is a social component' (Bloor 1976:13). And some philosophers, for example Mary Hesse, are sympathetic. She comes to the same conclusion as Bloor in her new work *Revolutions and Reconstructions in the Philosophy of Science.*[3] Sociological explanations will have to be given since the evidence, Hesse claims, will not determine one theory as being rationally preferable to its rivals.

The work on incommensurability by Kuhn, Feyerabend, and Wittgenstein has had a great deal of influence on strong programmers. In his new book on Kuhn, Barry Barnes remarks, 'Nothing in the nature of things, or the nature of language, or the nature of past usage, determines how we employ, or correctly employ our terms.' Moreover 'if nothing external determines what concepts are to refer to, then nothing external determines the truth or falsity of verbal statements' (Barnes 1982:30). The determining factors in how we use our concepts, and especially how we extend them, Barnes claims, are social factors.

The next three chapters are devoted to critically examining these claims in some detail. Chapter two will be (mainly) a look at Bloor's 'scientific' approach to science; chapter three will be (mainly) an examination of Barnes's account of language and related themes, which he calls 'finitism', and some of the sociological consequences he draws from it.

SOME EXAMPLES

Yet another factor in the sociological turn is the perceived successes of the recent case studies. I have been describing some of the programmatic features of the Edinburgh school's view of science, but now, in order to clarify just what the recent sociological turn is, a few examples are in order. The last few years have seen a great number of new sociological accounts of old stories, and I shall summarize three of these below for handy reference. Brief descriptions of a few of the paradigmatic case studies will throw a lot of light on the new sociological enterprise. The ones which I have chosen to recount are from the collection of examples that the Edinburgh school members have either produced themselves or have repeatedly referred to in approving terms.

Clearly, case studies are important to the issue. Of course, no single case study will make or break either side, but most people concerned with these issues will readily agree that historical considerations have some role to play. Just what that role is, is hard to say. It cannot be straightforward since the proponents of the arationality principle can readily allow that there have been some instances of 'bad' science: consequently, there should be cases where the rationalist and the sociologist will likely concur on the cause (i.e., sociological) of the scientific belief in question. Whatever the evidential role these examples play, legitimate or not, there can be no doubt that they have been influential in discussions of the issue. Here for handy reference, are brief synopses of three of the most influential.

My uncritical expositions of these three case studies should not, of course, be taken as endorsements. Though the following case studies have been cited as evidence for the sociological account of science, I shall not offer detailed criticisms of them in this book. All of my criticisms of the strong programmers and other cognitive sociologists of science will be at the programmatic or method-ological level. The fourth and fifth chapters, however, do argue for a specific way of understanding the evidential role of individual case studies, and that will have an indirect bearing on these examples.

Forman on Weimar Culture and Causality

The scientists of the Weimar Republic, according to Paul Forman (1971), saw themselves as under attack. And their perceptions were correct, for they were indeed living in a hostile intellectual environment. Following the First World War and Germany's collapse the public was quite disillusioned with science and technology. The tenor of the times was mystical and anti-rational. Even the little bits of science that the public was interested in, such as relativity, were often used to support anti-rationalist causes. The mystical romantic public sentiments of the post-war German public were in direct opposition to the perceived spirit of science, which in turn was seen as mechanistic, rationalistic, and wedded to causality.

This sense of general intellectual crisis was epitomized by Spengler's *Decline of the West*. The content of Western mathematics and physics, according to Spengler, expresses the 'Faustian' nature of contemporary Western culture; and the essential ingredient is the *Kausalitätsprinzip*. Mathematics, physics, causality, and ration-ality are lumped together and linked to death. In opposition to this is the creative, the living, that which embraces the incomprehensible 'Destiny' (*Lebensphilosophie*). But Spengler also added that physics in his day had exhausted its possibilities; doubts were arising about its principal concepts. Salvation for Western science would come about, he thought, but only when that science returns to its spiritual home.

Various remarks from a number of prominent scientists who embrace or at least seriously allude to the popular sentiment are cited by Forman. Wien, von Mises, Weyl, Born, Sommerfeld, and others are noted as making significant concessions to the importance of 'spiritual values', 'the mystery of things', 'traditional German idealism', and so on. In fact, these concessions went on to such an extent that Forman calls it a 'capitulation to Spenglerism' (Forman 1971:55).

The then popular view of a crisis of culture precipitated the crises in the various sciences after the war, according to this account. As for physics in particular, 'The *possibility* of the crisis of the old quantum theory was', says Forman, 'dependent upon the physicists' own craving for crises, arising from participation in, and adaptation to, the Weimar intellectual milieu' (Forman 1971:62).

Forman characterizes all those who embraced a non-deterministic view as espousing it in a moralistic and almost religious fashion, and moreover, as doing so only after the most frivolous considerations. In short, there was nothing rational about it. He concludes:

suddenly deprived by a change in public values of the approbation and prestige which they had enjoyed before and during World War I, the German physicists were impelled to alter their ideology and even the content of their science in order to recover a favorable public image. In particular, many resolved that one way or another, they must rid themselves of the albatross of causality.

In support of this general interpretation I illustrated and emphasized the fact that the program of dispensing with causality in physics was, on the one hand, advanced quite suddenly *after* 1918 and, on the other hand, that it achieved a very substantial following among German physicists *before* it was 'justified' by the advent of a fundamentally acausal quantum mechanics. I contended, moreover, that the scientific context and content, the form and level of exposition, the social occasions and the chosen vehicles for publication of manifestoes against causality, all point inescapably to the conclusion that substantive problems in atomic physics played only a secondary role in the genesis of this acausal persuasion, that the most important factor was the social-intellectual pressure exerted upon the physicists as members of the German academic community.

(Forman 1971:109f.)

And, moreover, he adds,

Although a readiness to view atomic processes as involving a 'failure of causality' proved to be, and remains, a most fruitful approach, before the introduction of a rational acausal quantum mechanics the movement to dispense with causality expressed less a research program than a proposal to sacrifice physics, indeed the scientific enterprise, to the *Zeitgeist*.

(Forman 1971:112f.)

For a critique of Forman, see the recent replies of J. Hendry (1980) and of P. Kraft and P. Kroes (1984).

Shapin on the Edinburgh Phrenology Debates

A Viennese doctor, Franz Joseph Gall (1758–1828), was the founder of phrenology. Its main principles were that the organ of the mind is the brain; that the brain is composed of separate organs, each associated with a different mental faculty; and that the size of the organ is related to the power of the associated mental faculty.

The phrenologists encountered two hostile traditions in Edinburgh: one was Anatomy, the other was Philosophy (i.e., the philosophical tradition from Thomas Reid and Dugal Stewart to William Hamilton). The ensuing debates in the early years of the nineteenth century were particularly strenuous. Steven Shapin has examined these debates and has given a sociological account of the episode, which runs as follows (Shapin 1975).

The early decades of the nineteenth century saw a serious growth of tension between the social classes in Edinburgh. The mercantile classes began to resent and reject the values and privileges of the upper classes, the landed gentry, the lawyers, and the professionals. The emerging middle classes established a

14

number of institutions for their own purposes, including their own paper, *The Scotsman*, which, says Shapin 'was critical of the Tories, the University, the established Church, and what it saw as intellectual obscurantism'. And he adds, 'it was no coincidence that *The Scotsman* supported the phrenologists in their dispute with the moral philosophers' (Shapin 1975:224).

Shapin begins by noting the enormous enthusiasm for phrenology amongst the middle and working classes. Popular lectures on the subject were amazingly well attended; working-class institutions regularly offered courses on the subject and even gave it pride of place. The attitude of Edinburgh University, however, was quite different; phrenology was never taught there. Often *anti*-phrenology lectures were given at the University, but phrenologists were seldom or never allowed to reply. The membership of the elite Royal Society of Edinburgh and the membership of the Phrenological Society were virtually disjoint. And so, says Shapin, the phrenology supporters must be considered 'outsiders'.

Shapin further claims that:

British phrenology was a social reformist movement of the greatest significance. Combe (George Combe, the most prominent phrenologist in Edinburgh) and his circle vigorously, and to some extent successfully, agitated for penal reform, more enlightened treatment of the insane, the provison of scientific education for the working classes, the education of women, the modification of capital punishment laws and the rethinking of British Colonial Policy.

(Shapin 1975:232)

Shapin certainly admits that there was a 'technical debate' with all the trappings of regular science, that is, with argument, evidence, and so on; nevertheless 'to say there *was* a technical debate is not to say that it can be or was separated from the social conflict, nor that such a technical debate does not reflect social and institutional divisions' (Shapin 1975:234).

Shapin's account of the Edinburgh phrenology debates is part of a debate he himself had with G. Cantor (Cantor 1975 and 1975a). The Cantor–Shapin confrontation especially repays a close examination because it is a head-on clash of two diametrically opposed ways of understanding science. While Shapin accounts for the episode in terms of social factors, Cantor, who is much more of a traditional intellectual historian, looks to the evidence, the arguments, and the reasons which were adduced for each side.

Farley and Geison on the Politics of Spontaneous Generation

Louis Pasteur and Félix Pouchet had a famous debate over spontaneous generation which lasted for about five years during the middle of the nineteenth century. Contrary to traditional accounts of this famous episode in the history of science, John Farley and Gerald Geison 'believe that [their] re-examination of the Pasteur–Pouchet debate reveals the direct influence of extrinsic factors on the conceptual content of serious science' (Farley and Geison 1974:162).

In a nutshell, spontaneous generation is the doctrine that living organisms can arise independently without parents from either inorganic matter (abiogenesis) or from organic debris (heterogenesis). Typical historical accounts claim that Pouchet set out with preconceived ideas favouring spontaneous generation while Pasteur, with his flawless experimental technique, destroyed the doctrine by doing a series of conclusive tests. It is quite a different picture that Farley and Geison paint.

In the middle of the nineteenth century, France was a politically conservative country, indeed, even reactionary. Louis Napoleon had come to power in 1848 with the support of the Catholic Church, and religious and political issues were inseparable. Church and State faced the perceived common enemies: republicanism and atheism. Indeed, very often atheists, positivists, and materialists truly were opposed to both Church and State. One striking instance is Clémence Royer, who translated Darwin's *Origin of Species* in 1864. In her preface she explicitly attacked the Catholic Church, calling it corrupt, ignorant, and responsible for all societal ills. Generally, there were considerable social tensions and they were often reflected in people's attitudes to science.

The Pasteur–Pouchet debate can be said to begin with the appearance in 1859 of Pouchet's *Hétérogénie, ou traité de la génération spontanée*. Aware of the political climate in which he was working, Pouchet inserted in his book explicit disclaimers of atheism and a prolonged justification of the assertion that his account of the theory of spontaneous generation was perfectly compatible with orthodox science *and* orthodox religion. His version of the theory posited a 'plastic force' which was capable of organizing molecules in special ways and endowing them with vitality. It was an egg, not an adult organism, which was spontaneously generated in this way. In defending himself against possible religious challenges, Pouchet claimed that Scripture did not contradict his claim that God might be always creating life; there is no reason to think that God stopped after the sixth day.

Pasteur's background to the debate is curiously contradictory, according to Farley and Geison. On the one hand, Pasteur had done a great deal of work on fermentation. In his work he needed to face the issue of the origin of the organisms responsible for the process. Pasteur was arguing against any chemical theory of fermentation, and so he had to argue, obviously, that the organisms pre-existed and did not arise heterogenetically. His later attack on Pouchet would seem to be of a piece with his views on fermentation.

However, Pasteur also did a great deal of work in crystallography where he had become convinced that molecular asymmetry (which manifested itself in optical phenomena) was intimately connected to life. Pasteur speculated that the force which brought molecular asymmetry about, an asymmetric force, was a kind of ordinary physical force. All of this, of course, suggests that abiogenesis could occur under ordinary mechanistic circumstances. Indeed, in his laboratory Pasteur tried to 'imitate nature' and to 'introduce asymmetry into chemical phenomena' (Farley and Geison 1974:178). Concerning all of this, Farley and Geison remark:

In fact, of course, Pasteur did not succeed in creating asymmetry or life and temporarily abandoned these experiments. But he continued to believe that abiogenesis should be possible under some such experimental conditions. He thus came into the debate over spontaneous generation faced with a curious dilemma. On the one hand, his work on fermentation led him to discount the possibility of heterogenesis, while on the other his theoretical views on asymmetry and life led him not only to believe in the possibility of abiogenesis but actually to attempt such a feat experimentally. If it seems illogical simultaneously to believe that life can be produced artificially from inorganic elements but not from a rich organic soup, it is essential to recall that Pasteur reached this paradoxical position as the result of

17

two quite separate research problems and to emphasize the distinction in his mind between symmetric chemical influences and asymmetric physical forces. Nevertheless, and this is the central point, Pasteur could deny the possibility of spontaneous generation only by suppressing part of his own scientific beliefs.

(Farley and Geison 1974:78f.)

Pasteur's political views were in complete harmony with the orthodoxies of the Second Empire, which is to say he was very conservative. He was a strong supporter of Louis Napoleon, dedicating a book to the Emperor and another to the Empress, and in turn he benefited greatly from Imperial favour. He once ran for the senate as a conservative and as a champion of the established order. Pasteur, according to Farley and Geison, had a 'general preference for order and stability over free speech, civil liberty or even democracy, whose potential for anarchy and mediocrity he feared' (1974:187). It was this strong conservative political view that influenced Pasteur to take the scientific stance that he did. In conclusion, Farley and Gieson say:

Remarkably enough, we are led to a conclusion precisely the opposite of that usually attached to the Pasteur–Pouchet debate. For we are persuaded that external factors influenced Pasteur's research and scientific judgement more powerfully than they did the defeated Pouchet. Having formulated his version of spontaneous generation prior to the politically significant Darwinian controversy in France, Pouchet maintained his views with striking consistency in spite of their presumed threat to orthodox religious and political beliefs which he fully shared. By contrast, Pasteur's public posture on the issue seems to reveal a quite high degree of sensitivity to reigning socio-political orthodoxies.

(Farley and Geison 1974:197)

Needless to say, there are rival accounts. In particular, this case study has been criticized by Nils Roll-Hanson (1979), who claims that external factors counted for little in the debate.

There are a number of other examples which I could just as easily have given, for instance, the recent studies by T. Brown, D. MacKenzie, H. Collins, M. Mulkay, A. Pickering, T. Pinch, S. Schaffer or any of those appearing in Barnes and Shapin (1979).

(See the bibliography for details.) They are recommended as further good examples which illustrate the sociological approach to understanding science.

Throughout this book, when I need an illustration, I shall usually confine myself to choosing from the recent products. But, of course, the newer case studies are not the only ones. One of the most famous from an earlier generation is Boris Hessen's (1931) Marxist study of 'The Social and Economic Roots of Newton's *Principia*'. This actually raises an interesting question about where Marx himself would stand on these issues. Marx would appear to be entirely with the sociologists when he remarks that 'The mode of production of material life conditions the general process of social political and intellectual life. It is not the consciousness of men that determines their existence, but their social existence that determines their consciousness.' (1859, trans. 1970:20f.) But no sooner does Marx say this than he seems to endorse the possibility of sharply separating ideology from objective science.

> With the change of economic foundation the entire immense superstructure is more or less rapidly transformed. In considering such transformations the distinction should always be made between the material transformations of the economic conditions of production which can be determined with the precision of natural science, and the legal, political, religious, aesthetic or philosophic – in short, ideological forms in which men become conscious of this conflict and fight it out.
>
> (1970:21)

Thus, on the one hand, many elements of intellectual life are the result of material conditions; but on the other, Marxist economics itself is objective science and not just the product of working-class interests. It is no surprise that contemporary Marxists are to be found on both sides of present debates on the nature of Science.

Although its members are principally concerned with social causation, the Edinburgh school also permits psychological explanations as legitimate in principle. These might include reference to anything from neurological structure to Freudian mechanisms. One of my favourites, because it is so absurdly amusing, was produced by the philosopher J. O. Wisdom (1953). It is a psychoanalytic account of *The Unconscious Origins of Berkeley's Philosophy*. Berkeley is infamous for producing an extreme form of

19

empiricism in which the existence of all matter is denied. In order to give the curious reader a glimpse of Wisdom's account of why Berkeley denied the existence of matter, I shall string together several of his numbered psychoanalytic 'interpretations':

I. Berkeley felt that he had poison inside him, clogging and destroying him. . . . II. Matter symbolised poison in the external world. . . . IV. Berkeley feared turning God into poison. . . . XXIV . . . Matter = Poison. . . . XXVI. External poison was a defensive projection of internal poison, recognition of which would have been intolerable. . . . XXXIII. In infancy Berkeley felt some faeces to be extremely good. . . . XXXV. Berkeley's fear of poison was the fear that the intensity of his desire to incorporate good faeces would make him unable to avoid taking in bad faeces as well. . . . XXXVI. In infancy Berkeley found his bad faeces intolerable and felt without sufficient power to expel them. . . . XXXVII. Matter = power of external bad faeces = projected destructiveness. . . . XL. God cemented the world together or created it by means of pure faeces, acting always with regularity. . . . XLI. God, a pure creator who commanded unsullied power and possessed a pure cement in the form of good faeces for building the world, is a substitute for the father Berkeley valued; and the mathematicians and deists, creators of Matter and bad faeces, godlike enemies of God, were substitutes for the father Berkeley hated because of withholding from his son his faeces and great power of defecation . . .

Unless they explicitly embrace such an account of Berkeley's intellectual output, then, of course, we should not saddle the Edinburgh school with it. I include it in my stage setting only for comic relief.

A TURN FROM WHAT?

While the sociological turn has been pursued with vigour and has produced several interesting results, it is nevertheless important to ask, 'Just what is it a turn from?'

In *Changing Order* Harry Collins remarks that 'During the last decade sociologists, historians, and philosophers have begun to examine science as a cultural activity rather than as the locus of certain knowledge. . . . [T]his new perspective demystifies the role of scientific expertise' (Collins 1985:1). Elsewhere he further adds, '[M]ost philosophers of science work with some version of the "canonical model" . . . [which stresses] the infallibility of experimentally generated knowledge' (1985:144).

Is that what the sociological turn is a turn from: 'certain knowledge'? 'mystery'? 'infallibility'? One can only marvel that it took the sociologists so long to lead us away from the papal-encyclical model of science. Barry Barnes paints a similar picture of the rationalist view which he wants us to reject.

> One common conception of knowledge represents it as the product of contemplation. According to this account, knowledge is best achieved by disinterested individuals, passively perceiving some aspect of reality, and generating verbal descriptions to correspond to it. . . . Invalid descriptions . . . often . . . are the products of social interests which make it advantageous to misrepresent reality, or social restrictions upon the investigation of reality which make accurate perception of it impossible.
>
> (Barnes 1977:1)

Of course, these are only caricatures of how rationalists view the activity and the results of science. The labels 'rationalist' and 'sociologist' are convenient, but they hide a wide spectrum of views. Rationalists are agreed among themselves on little; yet I doubt that there is a single one (who is in any way prominent) who thinks science is 'infallible', produces 'certain knowledge', or is generated through the 'passive perception' or 'contemplation' of reality, or who thinks that the sole cause of mistakes in science is that of social factors. It is instructive to note that neither Barnes nor Collins cites examples of such rationalists – which is just as well since none exist. Karl Popper, for intance, is confident of only one thing: that all our theories are very likely false. No one could be further than he is from rejecting 'certain knowledge' or in thinking that science is 'infallible'. Imré Lakatos and Larry Laudan both think of the history of science as a history of trying out frameworks. These frameworks (which they call research programmes and research traditions, respectively) are not generated through 'passive perception', but are rather actively imposed upon nature; in so far as a framework does not fit, it is rejected and a new one tried in its place. For Richard Boyd and William Newton-Smith, theories are getting closer to the truth, but they are not true now. Present failures are not necessarily due to social factors (see the section on the arationality principle, p. 5), but simply result from the fact that science is very hard to do. Being rational doesn't mean always getting it right.

In light of this, some of us might be pardoned for remaining stick-in-the-muds about the role of 'evidence', 'good reasons', and so on, and in thinking them the real driving force behind the development of science. One can only wonder how much of the appeal of the sociological turn stems from a caricature of the rationalist position.

Chapter Two

THE SCIENCE OF SCIENCE

> If sociology could not be applied in a thorough-going way to
> scientific knowledge it would mean that science could not
> scientifically know itself.
>
> David Bloor

The desire to be scientific is one of the most powerful motivations
of the new sociology of knowledge. Indeed, the pro-science flavour
of many recent sociological accounts of the growth of knowledge
sharply distinguish them from some of their predecessors, which
were often out to 'expose' or debunk science. David Bloor is
unabashedly the champion of scientific culture: the proper way for
science to know itself scientifically. This, he claims, simply requires
us to apply the scientific principles of sociology in a thoroughgoing
way.

Bloor's programmatic writings have been extremely influential.
That in itself is enough to make them important and worthy of
serious consideration: but they are often genuinely insightful, and
always provocatively intriguing as well. The first chapter of his
book *Knowledge and Social Imagery* is something of a manifesto which
signals, endorses, and promotes the new sociological turn. It
champions the strong programme in the sociology of science, which
is Bloor's 'scientific' approach to science. I have already given a
brief description of this programme in the preceding chapter; now
it is time to critically examine this view and see what the strong
programme amounts to.

REASONS AND CAUSES

The first tenet of the strong programme is Bloor's principle of
causality:

23

the sociology of scientific knowledge . . . should be causal, that is concerned with the conditions which bring about belief or states of knowledge. Naturally there will be other causes apart from social ones which will cooperate in bringing about belief.

(Bloor 1976:4)

In including the rider about other types of causes which might bring about belief, Bloor has such things in mind as psychological causes, which might include anything from neuroses and bizarre toilet training to complicated neural structures. It is not that he endorses any of these specifically; it is rather that, in principle, these types of factors can be appealed to for explanatory purposes while so-called 'rational factors' cannot. Moreover, there might be yet other types of cause as well, such as genetic factors. According to sociobiology, some of our beliefs are the result of a Darwinian evolutionary process; we have various cognitive inclinations because they have survival value. And Chomsky's claim that a child's knowledge of grammar is innate, due to genetic programming, would also seem a thesis which satisfies the spirit of the principle of causality.

Sociological causes, however, are Bloor's chief concern; these, he thinks, are the primary driving force in the development of scientific beliefs. What has traditionally been called 'a good reason for believing' is the sort of thing Bloor wants to exclude from consideration. (But see the section 'A Sociology of Reasons?' on p. 32.) The principle of causality is set up to contrast causes with reasons, and the target of his principle is rational explanation. To explain a belief, that is, to give a 'scientific account' of it, is to cite a cause; it is not sufficient to cite a reason. This is Bloor's implicit claim, and it is based on an assumption which he shares with many others that reasons simply are not causes. It is not an assumption which he states explicitly; he merely takes it for granted. But is this assumption true?

There are a number of things one might have in mind in denying that reasons are causes. For example, one might mean to deny (1) that the only way to acquire a true belief is by reasoning, or (2) that rationalizations, after the fact, are causes of belief, or (3) that reasons are the same kind of thing that causes are, or finally, (4) that reasons have something to do with belief whenever non-cognitive interests are at work. Let us look briefly at each of these.[1]

24

We regularly encounter people who believe the right thing for the wrong reason. It is perfectly easy to imagine someone being physically tortured in such a way that certain kinds of beliefs are created which just happen to be true. Brainwashing, we might image, is an effective means for making people believe that $2+2=4$. We can all admit this without throwing over the traditional view of science; for to concede that instances of this way of producing beliefs might occasionally occur is not to admit that they occur regularly in science.

Similarly, we must admit that after-the-fact rationalizations are a commonplace in daily life. We should even concede that such *post hoc* justifications have played a role in the history of science. Two centuries ago, for example, considerable 'research' was done on Africans which concluded that their position in the biological scheme of things was 'inferior'. Slave-owners gladly used this to justify slavery. Even if we concede (just for the sake of the argument) that such beliefs were indeed rational in the eighteenth century, we can rest assured that slave-owners probably believed what they did because those beliefs (rational or not) were so useful. Their economic interests were well served by such theories.

The existence of such rationalizations is unfortunate; but to admit their occasional presence is not to say that all or even much scientific reasoning is of the *post hoc*, self-serving nature. It is important not to be naïve about science and its history. The image of the noble scientist single-mindedly pursuing the truth to the exclusion of all else, and working in complete isolation from the society that she lives in, is an image to be forsworn. But it would be equally naïve to see nothing but self-serving rationalizations at work everywhere. Neither picture does justice to the intricacies of intellectual life, and both must be abandoned.

I am quite prepared to concur with the denial that reasons are causes in the first two senses of the four I mentioned, but not in the second two senses. I now turn to the consideration of these latter two.

The question 'Are reasons a kind of cause?' cannot be settled by examining the issue at the common-sense linguistic level, but this level is a good place to get a perspective on matters. The interesting thing to note is that we can substitute the words 'reason' and 'cause' in a wide variety of contexts and not change the meaning. A why-question can equally well be answered by citing either a 'reason' or a 'cause'. For instance, either word will do in the blank space in this sentence: 'The stone fell to the ground, the —— being gravity.' There are subtle differences, though, and they should be noted. In the sentence 'The —— the fire was careless smoking', we can use either 'cause of' or 'reason for', but not 'cause for' or 'reason of'. Whether this marks a deep difference between reasons and causes or is merely a linguistic accident, I cannot say. But, *prima facie*, most common-sense considerations indicate a very close connection between reasons and causes, close enough, perhaps, to say that reasons *are* causes.

Commonsense considerations, however, are not going to make the case one way or the other. So let me turn now to one of the more serious arguments for distinguishing between reasons and causes. Perhaps considerations similar to the following had some influence on Bloor in the formulation of his views.

Are reasons a different kind of thing from causes? A common claim runs like this: To discover the cause of an action or event (including the action or event of adopting a belief) requires some theorizing, observing, and empirical testing, even if only at a very elementary level. Whatever is eventually cited as the cause has only a tentative status; it is a conjecture which could be overthrown. The epistemology of beliefs, on the other hand, seems quite different. People have privileged access to their own reasons. Just as we can be certain about being in pain, so we cannot be mistaken about our reasons. Consequently, this line of argument concludes, since our knowledge of reasons is certain while our knowledge of causes is only tentative, reasons must be quite a different kind of thing from causes. And since Bloor wants us to be scientific, which is to explain by citing causes, it follows that our explanations should not be in terms of reasons.

There are a couple of responses we could make to this argument. One is to simply say that some causes – namely, reasons – can be known with certainty. But there is another answer I prefer to give, and that is to deny that there is such a thing as privileged access to one's inner mental life. That is, we should deny that there is such a thing as certain knowledge of one's own reasons for action. A classic example will illustrate this. Newton, in all good faith, said, '*Hypotheses non fingo*'. Moreover, he cited this as his reason for acting as he did in constructing some of his beliefs. Most historians would now agree, however, that Newton did indeed frame hypotheses. He had good reasons for his actions, all right, but they were not the stated reasons that he, in good conscience, put forward. The upshot, then, is that our knowledge of a person's reasons for acting are just as tentative and conjectural as any other cause. The belief that we clearly see our reasons is part of the myth of the given, a doctrine which is now totally discredited. So reasons and causes cannot be distinguished in this way.

Now to the fourth and final point: Do reasons have anything to do with belief when non-cognitive interests are at work? Reasons are not used exclusively for coming to the *truth*. Just as causation does not imply error, so, the presence of reasons does not imply the veracity of the final result. When our aim is the truth, then reasoning can help to get there; but we have other aims as well, and reasoning can be just as helpful in these other cases. Implicit in most of the work of the recent sociological turn is the claim that when they have shown that interests have caused a belief, then that automatically excludes any possibility that reasons have played the relevant causal role. Nothing could be further from the truth.

Consider Forman's study of the Weimar scientists. Does it not run as follows? The Weimar scientists wished to bring about a certain state, namely, being highly regarded by the public. They realized that presenting quantum mechanics in a non-deterministic way would help to bring about this result. Thus, they presented quantum mechanics in non-deterministic dress.

Now if this is indeed the structure of Forman's case study, then it does appeal to the scientist's reasoning processes to explain the events. His appeal to this reasoning need not be explicit because it is a form of reasoning which we all readily recognize and concur with. In Forman's article the scientists have suppressed cognitive goals for social ones. Nevertheless, these scientists still must do some reasoning in order to get where they want to go. Whatever the goals, there are rules of rational procedure. In this case they are so simple that there is no need for Forman to mention them; the reader can fill them in subconsciously.

The importance of reasons in bringing about belief, even when the goals are non-cognitive, is dramatically illustrated in the case of Pascal's wager. What Pascal did was to give a very clever argument for why we should believe in God. It is totally different from all other arguments for God's existence, since, according to Pascal, we should believe, not for cognitive reasons, but rather for pragmatic reasons, that is, for concerns of a purely practical and self-serving nature.

Pascal argued like this: Either God exists or not; we do not know which. We have a choice: we can believe that God does exist (and act accordingly) or we can believe that he does not. Which belief should we adopt, assuming that if God exists, then he rewards the faithful and punishes the atheists? Pascal set up a decision matrix with the various pay-offs. For example, if we believe in God and God does exist, then the pay-off is an infinite reward. On the other hand, if God does not exist, believers have merely squandered one day a week in pointless ritual; their loss is a finite one. The four possibilities are indicated in the table below.

	God exists	God does not exist
Believe that God exists	Infinite gain	Finite loss
Believe God does not exist	Infinite loss	Finite gain

It is obvious what the right thing to do is. From the point of view of self-interest, one should adopt the belief that God exists. Any belief that is adopted is something of a gamble, but it is crystal clear which is the best bet. The belief that God exists is self-serving to the highest degree possible.

(In passing I should note that ultimately the argument won't work. We need only imagine the possible existence of a god who prefers to hide and who favours atheists and rewards them for their disbelief. Thus, we should be atheists because of the greater pay-off, if we are right. The fact that Pascal's argument is not ultimately cogent has no bearing on what I said above. It can still be causally efficacious in bringing about beliefs in those who do not see through it. I should also note that Pascal assumed that one could brainwash oneself, so to speak, into the kind of belief state that his wager argument required. Just living with monks would do the trick, he thought.)

Suppose that a lifelong atheist, upon reading Pascal's wager argument, becomes a believer. How are we to explain this; what caused this new belief? The answer is clear: We should appeal to interests, just as Bloor would have us do. An interest was the cause of the belief. But it is essential to cite the chain of reasoning involved in the wager argument. It was this reasoning process that caused the subject to see clearly just what her interests really were. Without that reasoning process the new belief, that God exists, might never have occurred. And so, it is perfectly correct to say that reasons caused the belief. Some causes are reasons.

Let me elaborate on this theme. I think it is clear from the foregoing that the proponent of social causation will have to concede that *pragmatic reasons* are indeed causes of belief. In some cases, Pascal's wager will have provided pragmatic reasons for, and been the cause of, belief in God's existence. However, this is perfectly compatible with the sociologists' claim that non-cognitive interests are responsible for the production of belief. To undermine this claim we shall have to push further.

Contrast pragmatic reasons for belief with what I shall call *evidential reasons*. An evidential reason, roughly, is what has traditionally been meant by 'the evidence' or 'a good reason'. It is the sort of thing which scientists normally offer in support of their theories. And it is also the sort of thing natural theologians would usually offer to have us believe in God, in contrast to Pascal's argument, which tells us how best to serve our own interests. (For now, ignore the question whether the alleged evidential reasons are very good ones in any particular case.) With this distinction between pragmatic and evidential reasons in mind, let us now run through some possible explanation schemas.

Consider the following explanation of why A believes that God exists:

It is in A's interest to believe that God exists
_____
∴ A believes that God exists

This explanation sketch has a suppressed premiss, which is: Anyone with such an interest will so believe. But, of course, this is preposterous. We all recognize that quite often people do not know what their interests are, and sometimes, even when they do, they cut off their noses to spite their faces. So this explanatory form will not do. Let us try another, which is more plausible:

A believes that it is in A's interest to do X.
Anyone with such a belief will do X.
X = believe that God exists.
_____
∴ A believes that God exists.

The second premiss just says that people act in accordance with their perceived interests. Strictly, this is false, but for the sake of the argument let us here assume it is true. (A more sophisticated account would include non-deterministic beliefs, then we could turn the explanation schema into a probabilistic or statistical one.)

Anyone who has studied Pascal's argument and found it cogent will satisfy the premiss:

A believes that it is in *A*'s interest to believe that God exists.

Is this belief, which is a belief about a belief, held for pragmatic reasons? The answer is surely, no. There is one and only one thing in *A*'s interest here: to get it *right.* The truth, above all else, serves *A*'s interests in this case: not the truth about God's existence, which is inaccessible, but rather the truth about the wager argument. The belief in God is caused by pragmatic reasons; but the belief that belief in God serves one's interests is not caused by pragmatic reasons. It is caused by evidential reasons. So even evidential reasons, therefore, can play a causal role in bringing about interest-serving beliefs.

By having reasons be causes we reap an extra reward, for we break what is, even for Bloor, an unwanted asymmetry. The strong programme would have it (at least implicitly) that scientific theorizing is just epiphenomenal; our beliefs are the effects of external causes, but are never themselves causes. Such one-way causality is reminiscent of Newton's absolute space, which had a causal influence on matter – it made material bodies move in straight lines – but was itself completely unaffected by matter. This was long thought problematic, and it is considered a virtue of General Relativity that space and matter causally interact. It seems to me similarly virtuous to reject the causal asymmetry of Bloor's epiphenomenalism and to embrace the full causal symmetry involved in making reasons be causes.

In conslusion we can say that reasons are causes. And it follows from this that to explain the belief by citing the reason for it is to give the cause of the belief. This in turn implies that rational explanations satisfy the demands of Bloor's principle of causality. *Rational explanations are causal explanations.* The causal principle, which is a fundamental principle of Bloor's strong programme, does not divide the rationalists from the sociologists of knowledge. When properly understood, the rationalist embraces it too.[2]

ANTHROPOLOGISTS IN THE LAB

A favourite metaphor of recent years is the anthropological one. The tools and techniques of anthropology, a pre-eminently respectable and scientific discipline, are to be applied to the specific subculture of science. Scientific society is as amenable to anthropological study as any other exotic cultural group. This approach is certainly preferable to the investigations done by typical a priori rationalists, it would be claimed, since a priori accounts of science are as useless as armchair speculations on, say, Samoan dietary habits or the religious practices of New Guinea. It is also preferable to letting scientists speak for themselves. To repeat the remark of Sandra Harding, 'To what other "community of natives" would we give the final word about the causes, consequences, and social meanings of their own beliefs and institutions?' (Harding 1986:39).

The claim just to be doing anthropology is made by several champions of the sociological turn. In its most explicit form it can be found in the influential book by Bruno Latour and Steve Woolgar (1979), called *Laboratory Life*. The unmistakable suggestion is that, unlike the armchair speculators (i.e., philosophers), the anthropologists are getting out into the field to see how science is actually done, and the result is a factual report of their empirical findings. Undoubtedly, the metaphor is powerful and appealing, but it conceals a major confusion.

Let me illustrate with a story. Suppose there are two anthropologists investigating the same primitive society. They both notice the following about the village they are studying: First, the relatively poor people live on one side of the village; the rich on the other. Second, there is a sudden change in the behaviour of one of the two groups; the members tend to interact less; they shun one another and there is much talk about being 'unclean'. This much the two anthropologists agree on, but their explanations for the sudden change of behaviour are quite different. One of them notices that the change of behaviour is confined to just one of the social classes, and so assigns a class origin to this behavioural difference. The other anthropologist, unlike the first, believes in a germ theory of disease, and conjectures that some sort of germ which is confined to one side of the village is causing much illness; the 'shunning' behaviour is the villagers' reaction to the presence of disease.

Here we have rival anthropological ('scientific', 'causal') explanations for the villagers' behaviour. What is the moral of this story? It is very simple: Rationalists are anthropologists of science, too; it is just that they have a different theory about what generally causes a scientist's behaviour, namely, they think that often good reasons and evidence rather than social forces are the determining influences.

The problem is not the confusion of causes with correlations; to say that would be to trivialize the difference in approach. (It would also be gratuitously insulting since sociologists are well trained to look out for that very kind of mistake.) Rather, the difference between these two types of anthropologist of science is that they are armed with quite different accounts of the causal structure of the society they are both studying. The big question is not: Should we be anthropologists of science? Rather, it is: Which anthropological account is the right one? Is it in the interests account or the reasons account? There is no question of only one being the scientific account.

A SOCIOLOGY OF REASONS?

Readers who have taken the sociological turn may be getting quite impatient by now. They will say, 'Of course reasons play a role; but the fact that they do itself needs a sociological explanation.' Here is one rather full statement of this line of thought:

> What counts as an 'evidencing reason' for a belief in one context will be seen as evidence for quite a different conclusion in another context. For example, was the fact that living matter appeared in Pouchet's laboratory preparations evidence for the spontaneous generation of life, or evidence of the incompetence of the experimenter, as Pasteur maintained? As historians of science have shown, different scientists drew different conclusions and took the evidence to point in different directions. This was possible because something is only evidence for something else when set in the context of assumptions which give it meaning – assumptions, for instance, about what is *a priori* probable or improbable. If, on religious and political grounds, there is a desire to maintain a sharp and symbolically useful distinction between matter and life, then Pouchet must have blundered rather than have made a fascinating discovery. These

were indeed the factors that conditioned the reception of his
work in the conservative France of the Second Empire.
'Evidencing reasons', then, are a prime target for sociological
inquiry and explanation. There is no question of the sociology of
knowledge being confined to causes *rather than* 'evidencing
reasons'. Its concern is precisely with causes as 'evidencing
reasons'.

(Barnes and Bloor 1982:28f.)

Their claim is that *how* evidence is used is itself subject to social
forces. The fact that Pasteur used the presence of living matter in
Pouchet's preparations one way (to discredit Pouchet), rather than
another (to vindicate him), reflected Pasteur's conservative politics.
Evidence is not efficacious in its own right, according to Barnes
and Bloor, it is just another tool to be used in the promotion of
various social ends.

There are two things a rationalist should say in reply. One is to
point out the presence of an equivocation. This difficulty is
relatively easy to dispose of. The other point, which is much more
problematic, has to do with rules of inference.

First to the equivocation. A rationalist may well say something
like this: 'The presence of living matter in Pouchet's preparations
was a reason for rejecting the spontaneous generation theory.' Let
us break this down into two propositions: E for the evidential
reason 'Living matter was present in Pouchet's preparations', and
C for the conclusion 'The spontaneous generation theory is wrong'.
When a rationalist cites a reason E for a conclusion C, what is
actually intended is this: There are background beliefs B such that
given B and E we can infer C. (I will get to the nature of this
inference shortly.) Of course, should the background theories be
different, say B', then the right thing to say might be: B' and E
implies not-C. That is, the proper thing to say would be: 'The
presence of living matter in Pouchet's preparations was a reason
for *accepting* the spontaneous generation theory.'

If we tried to use E as a reason by itself, then we would fail to
justify a belief either in C or in not-C. It would then perhaps be
quite proper to give a sociological account of why Pasteur used E
as a reason for C rather than as a reason for not-C. But when the
evidence E is tied to a specific context we (often) get a unique
outcome. If the background theory is B, then E is a reason for C; if

the background theory is B', then E is a reason for not-C. The problem stems from calling E by itself a reason, when actually it is the whole cluster, B and E, which is the real reason for the further belief C. (There are two ways a rationalist might describe the situation. (1) Evidence E is a reason for C in the context of background beliefs B, or (2) E and B are a reason (context-free) for C. The difference is merely terminological and I doubt whether one way of speaking is preferable to the other.)

Barnes and Bloor seem aware of this when they remark that 'something is evidence for something else only when set in a context of assumptions which give it meaning' (ibid.). However, their example suggests they do not understand things in the same way that I have just outlined. For them the 'context of assumptions' is not a cluster of background theories but a context of social forces. They provide a typical instance of what they have in mind by a context of assumptions when they remark that 'If, on religious and political grounds, there is a desire to maintain a sharp and symbolically useful distinction between matter and life, then Pouchet must have blundered' (ibid.). The context they have in mind is not the context of background scientific theories. Until it is, their talk of paying attention to 'evidencing reasons' is highly misleading. Their 'sociology of reasons' results from ignoring the background beliefs, that is, only by ignoring what makes a good reason a good reason in the first place.

Well, so much for the equivocation; let me turn now to the second point, which has to do with inference. I said that C can be inferred from B and E, in the example above. But what about inference patterns themselves? What are they? Can they be justified by the rationalist, or are they amenable to a sociological account?

Apart from ordinary deductive reasoning, the rules of inference are a constant subject of debate among rationalists. Among the candidates are: Mill's Methods, Carnap's confirmation functions, Bayes's theorem, Popper's falsificationism, Lakatos's research programmes, Laudan's research traditions, Glymour's bootstrapping, and many others. In each case the proposed rules specify the conditions under which we may or may not infer the conclusion C. To have a theory of scientific method is in large part just to specify such rules.

Though no rule of inference is definitely known to be right, this should not be cause for concern. Linguists have long argued over the rules of good grammar. No particular rules are known to be right, either, but we would not conclude that people do not follow any grammatical rules. In both cases the task of discovering those rules is very difficult, but not hopelessly Utopian.

How is a rule of inference justified? This is a point on which rationalists may seem vulnerable. They will fall back on the context to warrant some particular belief, but this in turn may suggest that there are 'forms of inference [which] can be shown to be rationally justified in an absolute and context-free sense' (Barnes and Bloor 1982:40). The place at which rationalists seem most vulnerable is on the issue of induction, and where they seem least vulnerable is with respect to deduction. Yet it is on this, their firmest ground, that rationalists have come under attack. Barnes and Bloor begin by citing Lewis Carroll's well-known discussion of the justification of *modus ponens*. The tortoise asks why he should believe Q when he both believes P and beliefs 'If P then Q'. Achilles tells him that if he believes P and he believes 'If P then Q' as well, then he should also believe Q. The tortoise replies that this is all well and good, but why should he believe Q? We see the regress coming. Achilles, of course, is trying to give a new rule which will justify *modus ponens*, but any application of the new rule presupposes *modus ponens*. If the tortoise does not understand and accept *modus ponens* from the start, then he will never understand or accept any rule of the same form which tries to explain or justify it.

The conclusion that Barnes and Bloor draw from this is that *modus ponens* in particular, and all deduction in general, is not justified. The fact that people accept *modus ponens* or any other form of inference is only accountable on social or other non-rational grounds. (While they are inclined to take a conventionalist attitude toward styles of inference, they acknowledge alternatives such as nativism. 'Such a move', they point out, 'gives no comfort to rationalism: epistemologically, to invoke neural structure is no better than to invoke social structure' (Barnes and Bloor 1982:44).

There is a short and sweet response to the Barnes and Bloor argument, and there are several other considerations which considerably lessen its force. Let me begin with these latter, and then give the knockdown reply.

For one thing, we must note the difference between harmless circles and vicious ones. It is not that Barnes and Bloor run them together, but that often the charge of circularity seems more serious than it really is. What Carroll has done in his Achilles and the tortoise argument is to show that there is no noncircular account of how we use *modus ponens*. Contrast this with other principles that have been used in the past which turned out to be viciously circular. The principle of abstraction from naïve set theory (i.e., every condition determines a set) leads to an outright absurdity, namely, Russell's paradox of the set of all sets which are not members of themselves. This set is a member of itself if and only if it is not a member of itself. Russell's paradox contains a vicious circle, but the circulatory involved in *modus ponens* and other accepted rules of inference is the harmless sort of circularity which in no way undermines the correctness of the principle itself.

Secondly, to suggest, as Barnes and Bloor do, that a belief is not justified when it has no explicit justification, is seriously wrong. There is a significant chunk of anyone's knowledge which is not propositional. It is a case of knowing *how* rather than knowing *that*. Sociologists, even more than philosophers, like to stress this tacit dimension in scientific practice. (See, for example, Collins 1985.) Perhaps we can appeal to this notion when it comes to accounting for some of our basic inference patterns. We know how to reason correctly and we are perfectly justified in how we proceed even though we cannot produce a verbal justification of the techniques involved. For instance, I drive a car reasonably well. If asked why I turned the wheel and lightly braked at a particular time – which was exactly the right thing to do in the circumstances – I would probably be rendered speechless, even though I would be perfectly justified in believing I did the right thing.

Now to the promised short and sweet reply to the Barnes and Bloor claim that logical truths are conventional. There are at most a finite number of distinct conventions, but there are infinitely many distinct theorems. It takes logic to derive these consequences from the conventions. Thus, at most some logic is true by convention; the rest is true by convention plus logic. (This argument is by now part of the logician's lore. It can be found in many places, for example Quine 1935.)

This reply to conventionalism is utterly devastating, but inevitably there are replies. One stems from Wittgenstein, or at least, from his followers. The 'radical conventionalist' says it is not just the axioms and rules of inference that are conventionally decided upon. Each individual theorem that we think of as an inevitable consequence of the axioms and rules of inference is itself a convention. The idea that there are infinitely many consequences already existing independently of us, is a mistake. There is no question of using (already given and understood) logic to derive consequences from conventional assumptions; that something is taken to be a consequence is itself conventional.

This gets around the objection to conventionalism. Apart from calling radical conventionalism names – 'ridiculous' comes to mind – I do not see how to answer it directly. But there is a crushing indirect answer. If this radical conventionalism is true, then sociology of knowledge is impossible. This is because conventionally established rules can no longer serve interests, since doing anything can count as obeying those rules. Pity the poor despot who makes tax laws with an eye to lining his own pocket. A population which pays not a penny in taxes can rightly claim to have faithfully obeyed the letter and spirit of the law.

IMPARTIALITY

So far I have been concerned with the first tenet of the strong programme, the causality principle, and its many aspects and ramifications. Now it is time to move on to the next. The second principle of the strong programme is known as *the principle of impartiality*. It says that a proper account of science 'would be impartial with respet to truth and falsity, rationality or irrationality, success or failure. Both sides of these dichotomies will require explanation' (Bloor 1976:4).

Bloor is absolutely right about this. And we need not worry about how to interpret this principle, as we did for the principle of causality, in order to find it acceptable to the rationalist. Every question of the form: 'Why did Newton believe P?' deserves an answer; and it matters not whether P is true or false, rational or irrational, successful or unsuccessful. Bloor and the rationalist are of one mind on this question.

Why does Bloor feel obliged to assert this principle, especially when it is so obviously true? He is upset by philosophers and historians who make a distinction between 'internal' and 'external' history of science. The former of this pair is the sequence of events driven by rational scientific factors, while the latter is the set of events which are determined by non-scientific factors; they are rational and non-rational beliefs, respectively. People who make this distinction between the internal and the external usually supplement it with the requirement that evidence and good reasons be cited to explain the one, while political, social, and psychological factors be employed to account for the other. This is quite different from requiring that no cause be given for true or rational beliefs. Everything is explained (or at least admitted to be in need of explanation), though different factors are cited in the two cases.

But is this distinction between the rational internal history of science and the non-rational external history, with their corresponding different styles of explanation, a legitimate distinction? This is really the crucial question, and Bloor's negative answer is embodied in his next principle.

THE SYMMETRY PRINCIPLE

The heart of the strong programme, and of any cognitive sociology of knowledge, is embodied in the symmetry principle. This is the core assumption of the new sociological turn. It is the claim that a proper account of science 'would be symmetrical in its style of explanation. The same types of cause would explain, say, true and false beliefs [and rational and irrational beliefs]'[3] (Bloor 1976:5).

The symmetry principle is directed against those who would insist that some kinds of beliefs should receive an entirely different kind of explanation from other sorts of beliefs. For instance, the traditional view of the history of ideas calls for 'good reasons' to be cited as the cause of any rational belief, while social causes are only appealed to in the case of non-rational belief. This is just the arationality principle discussed above. But Bloor wants only one kind of cause to be appealed to in *every* case, that is, in both the rational/internal and the non-rational/external case. Elaborating on his symmetry principle, he writes:

If these theories are to satisfy the requirement of maximum generality they will have to apply to both true and false beliefs, and as far as possible the same type of explanation will have to apply in both cases. The aim of physiology is to explain the organism in health and disease; the aim of mechanics is to understand machines which work and machines which fail; bridges which stand as well as those which fall. Similarly the sociologist seeks theories which explain the beliefs which are in fact found, regardless of how the investigator evaluates them.

(Bloor 1976:3)

At a certain level of generality, Bloor's maxim sounds very plausible. Physiologists are indeed concerned with the healthy body, just as with the sick one; and physicists interested in the strengths of materials are as concerned with standing bridges as with those which have collapsed. So what possible justification could there be for a divison of labour which calls on the sociologist only when beliefs are flawed? Surely they have as legitimate a concern with good theories as with bad ones. Bloor's line of thought may seem warranted at first blush, but we need not pursue it very far to see the absurdities it leads to.

What does it mean to give 'the same type of explanation'? Do bridge-builders really give the same type of explanation for why one bridge is standing as for why a second has collapsed? The presence of an earthquake will explain why a bridge has fallen; but we certainly wouldn't invoke the existence of an earthquake to explain why a standing bridge is still standing. Similarly, we might call on the germ theory to explain why someone is running a fever, but we don't appeal to the presence of germs to account for a normal temperature. Bloor is left with an absurd situation if he means that all explanations are to be *identically* the same. However, it would be very uncharitable to read him this way, so let's try a more liberal interpretation of his expression.

Suppose that by 'the same type of explanation', Bloor is merely advocating scientific (by which he means causal) explanations for all events: for bridges which are standing as well as fallen ones, for beliefs which are true/rational as well as false/irrational, and so on. Presumably this would just come to advocating causal explanations for everything. Well, as we have seen already, both sides agree with this. The rationalist will cheerfully give 'the same type of explanation' for all beliefs if this just means giving causal explanations for them; though, of course, the rationalist will pointedly add: 'Some causes are reasons'.

The first way the symmetry principle was interpreted led to an absurd outcome; the second to a view so vacuous that all would embrace it. Is there anything between these two? Is there a version of the symmetry principle which is plausible and which separates the rationalist from the sociologist? Let us try once more.

Perhaps Bloor would find something like this congenial: Whenever a given discipline (say, physics, physiology, sociology) provides an explanation for a phenomenon P, when P is the case, then it must also provide an explanation for not-P, should not-P be the case. That is, if a theory is prepared to account for P, then, by the symmetry principle, it must also be prepared to use its own resources to account for not-P when it is not-P which needs explaining.

So, for example, if physics is to explain why a bridge is standing then physics should also explain why a second bridge has fallen. The whole range of bridge posture is in the domain of physics, while psychology, astronomy, and literary criticism are forever indifferent to spanning structures. It is not that an identical explanation is called for each time, but rather that the same principles, the same concepts, must be invoked in both cases. Masses, forces, stresses, etc., will be appealed to to account for the phenomena. There will be different masses, different forces, and different stresses for standing and for collapsed bridges respectively, but it will be the presence of forces, stresses, etc., which in each case cause the phenomenon in question.

Similarly, given that we rightly appeal to, say, class interest to explain someone's belief in the divine right of kings, then, by the symmetry principle, we should also appeal to class interests (though they will be different interests) to explain why some other person is a republican.

This way of putting the symmetry principle makes it vastly more plausible than the two preceding ways. And at first blush it also seems to cut deeply against the rationalist programme. It is conceded by rationalists that sociological factors are the causes of irrational beliefs. But on this version of the symmetry principle, to allow that the sociologist has the right explanation for the irrational belief that P, is to allow that the sociologist will also have the right explanation for the rational belief that not-P. Plausible as this seems, however, there is an important consideration which completely undermines it.

Theories never explain phenomena by themselves. They require the help of auxiliary theories and initial conditions. Newton's theory of gravitation, for instance, does not explain the observed motion of Mars. To get it to do that we must add an initial condition, where Mars was to start with, and we need optical theories to tell us what we are seeing (i.e., the apparent as opposed to the true position of Mars, and so on). Is the observed motion of Mars then a phenomenon in the domain of gravitation theory or in the domain of optics? There is no obvious right answer; it is the combination of factors which account for the observed motion of the planet. Perhaps it is correct to say that it is in the domain of every theory which figured in its explanation. (Laudan (1976), for example, makes this claim.)

Supposing this to be so, what then is the domain of the phenomena of belief? Since sociological theories sometimes figure in the explanation of belief, Bloor might like to say that belief *always* belongs in that domain. But we can also say this: Often good reasons provide an explanation for belief, so reasons should always be used. And this is certainly not implausible, for even when interests are at work, reasons play a causal role in achieving non-cognitive ends, as we saw in the Pascal wager example. Thus, belief is entirely in the domain of the theory of rationality; reasons provide the right sort of causal explanation for all beliefs.

Since theories explain only with the help of auxiliary theories we still have the possibility of a sharp difference between rational and irrational beliefs. The auxiliary theory in each of these two cases might be a statement of the agent's goals. We can call a belief rational when the agent, first, has cognitive goals (i.e., when she seeks the truth or empirical adequacy, for example), and second, has good reasons for the belief. We might still call a belief irrational even though the agent has very good reasons for it. This can occur when, as in the Pascal wager example, the agent simply has other goals than cognitive ones and, instead of truth, seeks eternal bliss.

There is much more to be said on this subject; for now, however, I think we can safely conclude with this: there is no way of interpreting the symmetry principle which gives comfort to the sociologist of knowledge.

REFLEXIVITY

It is a commonplace to see doctrines such as Marxism and scepticism turned against themselves. 'Marx says that all beliefs are ideological, thus, Marxism must be ideology too.' 'The sceptic

41

says that no belief is trustworthy, so, we shouldn't trust the sceptic either.' Such 'self-refuting' objections are usually unfair – they are certainly unfair in the cases of Marx and the sceptic. But they do have to be met by any theory which looks in any way vulnerable to the charge, and Bloor's theory certainly looks vulnerable. His way of meeting the self-refutation challenge is to embrace the *principle of reflexivity*. A proper account of science

> would be reflexive. In principle its patterns of explanation would have to be applicable to sociology itself. Like the requirement of symmetry this is a response to the need to seek for general explanations. It is an obvious requirement of principle because otherwise sociology would be a standing refutation of its own theories.
>
> (Bloor 1976:5)

In his elaboration of this principle, Bloor remarks that it is simply false to equate causation with error. And he is quite right. A brick might fall on someone's head, causing the person to believe that the universe of general relativity is a four-dimensional differentiable manifold whereas he had previously thought it was a poached egg. Or, to take a more ordinary example, perhaps mercantile interests caused the shift in belief from the Ptolemaic theory to the much better Copernican one. Self-referring is one thing; self-refuting is quite another. Bloor's strong programme, he claims, is the first, but not the second.

All of this, unfortunately, misses the point even though it is quite right. The question is not whether Bloor's theory is true and harmlessly self-referring, or on the other hand, self-refuting and hence false. Rather, the question is whether we should believe his theory. Bloor's claim is that it is not evidence, but instead social factors, which cause belief. If Bloor is right and if he wants us to believe him, then he must drop bricks on our heads or alter our class interests or some such thing. There is no point in arguing his case; for if he is right, then arguments must be causally ineffective. If we have similar interests to Bloor, then, perhaps, we will have similar beliefs; and should our interests be opposed, then our beliefs could be expected to be contrary, too. All of this will occur quite independently of Bloor's argument. If *Knowledge and Social Imagery* is right, then it is destined to have no direct impact on intellectual life. But since it has had an impact, it must be false.

MOTIVES?

The principle of reflexivity invites speculation on the question: Why do people believe the strong programme? There should be, according to Bloor's principle, some sociological cause of this. Rationalists, by and large, will even agree on this point since they will likely see the strong programme as irrational and they further think that irrational beliefs are in need of some sociological explanation. Ironically, this may be the one area where we get real agreement between the two sides.

Loren Graham has offered a specific account of Hessen's famous study, 'The Socio-Economic Roots of Newton's *Principia*'. According to Graham, the time when this study was done, 1931, was

> a time of great political and economic stress . . . Soviet physics and Hessen personally were under very unusual pressures. Hessen's main concern in previous months had been to protect Einstein's relativity theory from attacks by vulgar Marxist ideologists. Hessen's paper on Newton was carefully crafted to support this defensive effort and simultaneously was aimed at strengthening Hessen's own political situation.
>
> (Graham 1985:705)

The social climate that contemporary Western academics work in is, of course, quite different from the one that Hessen worked in. So what makes some of them adopt the strong-programme approach to science? What might the cause be? Could the answer lie along political and social lines? Barnes and Bloor do not shy away from such speculations about the motives of their rationalist rivals. First they raise the question:

> [Though] the relativist must retire . . . to gaze from some far hilltop on the celebratory rites of the Cult of Rationalism, he can nevertheless quietly ask himself: what local, contingent causes might acount for the remarkable intensity of the Faith in Reason peculiar to the Cult?

They continue with this speculation:

> A plausible hypothesis is that relativism is disliked because so many academics see it as a dampener on their moralizing. A dualist idiom, with its demarcations, contrasts, rankings and

evaluations is easily adapted to the tasks of political propaganda and self-congratulatory polemic. This is the enterprise that relativists threaten, not science.

(Barnes and Bloor 1982:47)

Rationalists, will of course, simply dismiss this and instead hold that it is rational to be rational – good reasons are the cause of their beliefs. But what, we may continue to wonder, causes belief in the strong programme? A plausible view is that many who embrace the sociological turn are liberal-minded, with a strong sense of egalitarianism and tolerance for diverse cultures. People with such political sentiments are often ethical relativists. Fearing that they might otherwise be standing in moral judgement of other cultures, the liberal-minded will often extend their ethical relativism to cover *all* beliefs, even those we call scientific. Cognitive relativism stems from a desire on the part of its proponents to have a theory that coheres with the moral sentiment of tolerance. Indeed, Bloor quite explicitly wants a unified view: 'Philosophers sometimes perplex themselves because moral relativism seems philosophically accept-able but cognitive relativism does not. . . . Scientifically, the same attitude towards both morality and cognition is possible and desirable' (Bloor 1976:142).

I can only speculate on this question as I have little in the way of evidence, but evidence is not entirely lacking. Consider what Barnes says in arguing for an instrumental view of language:

in implying that the common-sense term [T] is ontologically inadequate, it sets the expertise of a subculture above the common-sense knowledge available to the lay person. Conversely, peaceful co-existence of [scientific T] and [T] usually implies peaceful co-existence between a subculture and the wider community, with the former either having no wish to, or no power to challenge the latter. Although there is no necessary connection here, realism does tend to be the language of cognitive colonialism and instrumentalism the language of compromise and toleration.

(Barnes 1981:328)

And as a further example consider Farley and Geison, who quite explicitly worry about reflexivity:

In writing this paper, we have quite naturally considered the

possible influence of external factors on our own interpretation of
the spontaneous generation debate. . . . Especially because we
find distasteful many of Pasteur's religious, political and
personal attitudes, our interpretation may well differ from that of
historians and scientists of more conservative persuasions.

(Farley and Geison 1974:198)

If I am right in my speculation, then the political sentiments of
some sociologists, which are quite laudable in themselves, have led
them astray. This is not the place to launch into a discussion of
why ethical relativism is a confused doctrine; the appropriate
distinctions which need to be drawn can be found in almost any
textbook on ethics (see, for example, Brant 1959). Contrary to
what Bloor suggests when he calls for a united account of ethical
and cognitive relativism, there are very few philosophers who are
moral relativists or who find the doctrine in any way acceptable.
Bloor and others have simply confused ethical relativism with the
(arguably) absolutely good ethical characteristic of tolerance. I
agree with Bloor that a unified view of ethics and science is both
desirable and possible, but neither is relativistic.

Of course, there is a wide spectrum of political opinion among
sociologists of science; I am not so naïve as to think my explanation
applies to all.[4] I only think it applies to a significant number.

Finally, let me add a word about the nature of this speculation.
It is not of the *ad hominem* variety, as speculations about motives
often are. Confused thinking about ethical relativism can sustain a
belief in cognitive relativism. To point out that the latter belief
stems from the former legitimately undermines that belief. It would
not be legitimate to say (even if true, which I am sure it isn't) that
belief in the strong programme is sustained by the desire to gain
notoriety, to get rich, or to promote fascism. The important
difference is this: if the desire for notoriety is the cause of the belief,
that desire still remains after it has been pointed out. But if
confused thinking about ethical relativism is the sustaining cause,
then pointing it out is the first step in having the whole situation
rectified.

CORRELATIONS AND COINCIDENCES

A principle which Bloor undoubtedly accepts but does not
specifically mention in his 'scientific' justification of a sociological

approach is *the principle of the common cause*. It could turn out to provide the most powerful argument in the strong programme's arsenal. The principle is a very simple one: significant correlations have common causes. If all the lights go out at once, we assume it is not a coincidence but that there must be a common cause of the sudden darkness, such as a blown fuse. The reason this principle is significant for the sociologist's case is that often there are correlations between the sets of people holding theories and various social conditions. Steven Shapin recognizes this and insists on the principle of the common cause as something to which the historian of science must pay attention.

> historical explanation should, and very often does, aim to reduce the domain of the 'coincidental' by searching out links, parallels and connections between one existential factor and another, between existential factors and thought, between one sphere of thought and another. Historians, and not just sociologists, often function as if it is their business to search out such parallels, to cherish them, to attempt to make sense of them when revealed by making every effort to weave them into an integrated narrative. Thus, if it is revealed to us Huttonian geologists, say, tended to be Whigs while Wernerians tended to toryism, we do not discard this information, we do not cast it adrift in a footnote or an aside; rather we recognize a parallel of this type as the very stuff of history, as a challenge to our capacity for integrative thought. In other words, as historians, we attempt to reduce coincidence in our materials. So one should look for social differences between people maintaining one intellectual viewpoint and another; having found them, we are obliged to make sense of them.

(Shapin 1975:221)

There is no doubt that the principle of the common cause is an important working rule of both science and everyday life. Shapin is absolutely right to insist that the interpreter of science must obey the principle, too. Correlations must always be treated with some caution, however, since often we fail to get the right reference class, as in this example:

> *Life* magazine reported that all fifteen members of the choir of a church in Beatrice, Nebraska, due at a choir practice at 7:20 p.m., were late on the evening of 1 March 1950. The

minister and his wife and daughter had one reason (his wife delayed to iron the daughter's dress); one girl waited to finish a geometry problem; one couldn't start her car; another couldn't start her car; two lingered to hear the end of an especially exciting radio program; one mother and daughter were late because the mother had to call the daughter twice to wake her from a nap, and so on. The reasons seemed rather ordinary, but there were ten separate and quite unconnected reasons for the lateness of the fifteen persons. It was rather fortunate that none of the fifteen arrived on time at 7:20 p.m., for at 7:25 p.m. the church building was destroyed in an explosion. The members of the choir, *Life* reported, wondered if their delay was 'an act of God'.

(Weaver 1977:212)

If we looked to the set of *all* exploding churches, we would not, I dare say, be tempted to postulate a common cause, an intervening god, who saves the day.

There are other cautions to be taken as well. Suppose A and B are correlated, for instance. The principle of the common cause can be evoked in three different ways: (1) A is the cause of B, (2) B is the cause of A, or (3) there is a C which is the cause of both A and B. In Shapin's version of the Edinburgh phrenology debates (1975) he claims the existence of correlations between phrenologists and the middle classes and between anti-phrenologists and the upper classes. There are, accordingly, three different ways each that these two correlations could be accounted for using the general principle. I suppose we can reasonably rule out that believing in phrenology (anti-phrenology) caused one to become middle-class (upper-class). For Shapin, the causal connection runs the other way: being middle-class (upper-class) causes one to believe in phrenology (anti-phrenology) though, of course, this is putting it crudely.

So here we have a direct tie between a social factor and a cognitive factor. One's social class seems to be causally connected to one's beliefs. Shapin postulates the direct connection between social factor and belief, but the direct connection is not the only one which satisfies the principle of the common cause.

There is an alternative way to view this situation. Perhaps things were more like this: Being upper-class gave one a significant amount of leisure time. This gave rise to the pursuit of art and science, etc., if for no other purpose than the relief of boredom. Among the disinterested beliefs adopted were certain anti-phrenology views. Here is an account which causally connects being upper-class with being anti-phrenology, but it is not an account which lets being upper-class have anything to do with the content of the belief.

It is not my intent here to challenge sociologists' factual claims in their case studies. It is only to show that different conceptualizations are possible. There are subtleties involved in the principle of the common cause which cannot be ignored, and we should not always jump to the conclusion that the direct causal connection between A and B is the one responsible for the correlation, when one does exist. Nevertheless, this does not diminish the fact that the principle does, I think, ground a significant intrusion of the social into the very content of science. Shapin is absolutely right to insist upon it. Other attempts to ground the strong programme on 'scientific' principles failed, but this principle has at least the potential to go some way in forcing a serious sociology of knowledge upon us. In the last two chapters of this book I shall take up this issue again.

Chapter Three

FINITISM

There are a number of related doctrines which all have to do with the finite nature of our experience. Theories, for instance, have infinitely many different observable consequences, yet we must evaluate them on the basis of only a finite number of empirical tests. And concepts, to cite another example, are learned with only a finite number of exemplifications: as children we learn what 'apple' means by being presented with a few examples. Yet the full extension of *apple* or any other concept, we commonly suppose, includes indefinitely many more instances than the few we have encountered. These simple considerations suggest room for social causation.

The arguments typically run like this: first, if theories really are *underdetermined* by the data, then the choice of one of them over any other (which is equally consistent with the finite data available) cannot be rational, but must instead reflect social interests (or some other non-rational cause). And secondly, suppose the actual extension of a concept is really no larger than the finite number of examples used ostensively to define it. Then whenever some newly encountered object is judged as being within or as being without the extension of a given concept, it must be viewed as a conventional decision reflecting social interests. Though briefly put, these are the ways a case for social causation is often formulated.

Underdetermination, incommensurability, and conventionalism, are common finitist themes. What I shall do in this chapter is to examine some aspects of finitism and see to what extent they vindicate a serious cognitive sociology of knowledge.

UNDERDETERMINATION

A geometric metaphor nicely illustrates the problem of underdetermination of theories by data. Suppose we have a graph with a finite number of points marked on it. How many different (possibly curved) lines can be drawn on the graph which connect all of the points? Answer: infinitely many. Analogously, how many different theories will correctly account for the finite amount of data that we have at present? The answer is the same: infinitely many. But if this is the case, why, then, do scientists choose one theory over any of its rivals? If there are no data which single out one of the theories as being uniquely compatible with it, then the choice cannot be a rational one, or so it would seem.

To make a rational choice is to decide on the basis of the available evidence. But here there is no available evidence that makes a difference. So the cause of the choice must have been something else such as a social force, and it is the social which sustains the acceptance of any theory. Thus Bloor remarks:

> It is largely a theoretical vision of the world that, at any given time, scientists may be said to know. It is largely to their theories that scientists must repair when asked what they can tell us about the world. But theories and theoretical knowledge are not things which are given in our experience. They are what give meaning to experience by offering a story about what underlies, connects and accounts for it. This does not mean that theory is unresponsive to experience. It is, but it is not given along with the experience it explains, nor is it uniquely supported by it. Another agency apart from the physical world is required to guide and support this component of knowledge. The theoretical component of knowledge is a social component . . .
>
> (Bloor 1976:12f.)

Michael Mulkay is of a similar turn of mind, citing philosophers for support. 'Sociologists and philosophers have converged on a conception of science as an interpretive enterprise, in the course of which the nature of the physical world is socially constructed' (Mulkay 1979:95). A social interpretation of science is all that is possible, he claims:

> The indeterminacy of scientific criteria, the inconclusive

character of the general knowledge-claims of science, the dependence of such claims on the available symbolic resources all indicate that the physical world could be analyzed perfectly adequately by means of language and presuppositions quite different from those employed in the modern scientific community. *There is, therefore, nothing in the physical world which uniquely determines the conclusions of that community.* It is, of course, self-evident that the external world exerts constraint on the conclusions of science. But this constraint operates through the meanings created by scientists in their attempts to interpret the world. These meanings, as we have seen, are inherently inconclusive, continually revised and partly dependent on the social context in which the interpretation occurs. If this view, central to the new philosophy of science, is accepted, there is no alternative but to regard the products of science as social constructions like all other cultural products.

(Mulkay 1979:61)

The great problem with arguments like this is that they play on the finitude of our cognitive life only in so far as it is convenient to do so. It is true that there are infinitely many different theories all compatible with the data that we have. But in the same sense that there are an infinite number of theories, there are also an infinite number of data. This is the logical or platonic sense. However, our real choice is always limited to a very few theories; what we actually have to deal with is only a finite number of alternatives.

It is only by blurring the logical sense of 'exists' (in which there do exist infinitely many different theories) with the available-at-hand sense of 'exists' (in which only a very few theories exist) that the argument from underdetermination gets off the ground.

Only the naïve among us see scientists choosing the true theory on the basis of the evidence. What scientific rationality, if it exists at all, would consist in is choosing the best theory from available rivals on the basis of available evidence. Historically, the actual choices have been, for instance, between Ptolemy and Copernicus, between Newton and Einstein, between classical physics and quantum physics. In each case the actual number of choices available was small indeed. And in almost every case the data did single out one theory as being better than its available rivals.

Real empirical equivalents do arise from time to time. So what

51

happens then? What is the attitude of the scientific community when real underdetermination occurs?

Consider Poincaré's views on the geometry of space (Poincaré 1952). Physical space, he claims, can be said to have some geometrical structure or other (i.e., some Euclidean or non-Euclidean geometry) only relative to a particular way of measuring lengths. Before a method of measuring is chosen, however, space has *no* geometry. There is no fact of the matter, for instance, about measuring rods as to whether they do or do not change their lengths as they are moved about in space. It is arbitrary which we choose. If we declare that measuring rods do not change length as they are moved about, then we will 'discover' that space is, say, non-Euclidean; and if we declare that measuring rods do change their length (in some complicated way to be specified) then we will 'discover' that our physical space is Euclidean. The geometrical structure of space is a convention, since it depends upon our arbitrary stipulation of how a measuring rod behaves.

Poincaré's views have been the subject of much controversy and, needless to say, he has many supporters and critics.[1] But that is not important here. What matters is the consequent attitude scientists take when faced with an example of what they believe is real underdetermination. Poincaré thought that all possible data equally confirmed any geometry. So he thought the choice was a convention, and he made this explicit. The existence of explicit conventional choices does not help the cause of the sociologists of knowledge; it hinders it. Poincaré opted for Euclidean geometry because he thought it was easier to use: he certainly did not think it was true. Did this in some way reflect his interests? Of course, he thought Euclidean geometry served his pragmatic interests since it was more practical, easier to use. But this doesn't serve any further *social* interest.

It might have been in the interests of the Weimar scientists to espouse indeterminism, but they would have completely undermined their own cause if they had added that their espoused views were merely the result of a conventional decision. It might be in my interests to get people to start believing in the Greek myths again; but I would destroy myself if I added, 'Oh, by the way, all this stuff about Zeus is just a myth.' Poincaré's espousal of Euclidean geometry would serve his non-cognitive interests only in so far as he can get others to think it is true; and this Poincaré disavows in the strongest possible terms. Conventionalism, when it is explicit, is no help to the anti-rationalist.

Thus far I have treated the problem of underdetermination as if the situation assumed theories T and T', which both did equal justice to observations O. What this presupposes is a neutral observation language, as it is often called. The existence of such theory-independent reports of experience is quite problematic. Recent philosophical work, especially that of Hanson (1958), Kuhn (1970), Feyerabend (1975), Hesse (1980) and others has shown rather persuasively that *observation is theory-laden*. We see and we describe only through the eyes of a theory.

This theory-ladenness leads to one aspect of the problem of *incommensurability*. Rather than having a straightforward situation of two theories, T and T', implying, say, the two observation statements O and not-O respectively (where we could then check whether O obtained and so decide between the two theories), we have instead T and T', implying O and O'. They talk past rather than contradict one another.

If it was already a problem that theory choice is based on only a finite amount of data, then the problem is now compounded by the further fact of not having common neutral data. How data are viewed will depend on the theory being tested.

How extensive theory-ladenness may be is a matter of controversy. Some think it confined to the observation language, while others, especially Feyerabend, think that theory-ladenness is all-pervasive and even conditions one's norms of rationality. The latter version is implausible, but here is not the place to delve into a long investigation of the problem since it will be taken up below. The more plausible versions of theory-ladenness do not present an insurmountable problem for rational theory choice, as a nice example developed by Mary Hesse illustrates:

Anaximenes and Aristotle are devising a crucial experiment to decide between their respective theories of free fall. Anaximenes holds that the earth is disc-shaped and suspended in a non-isotropic universe in which there is a preferred direction of fall, namely the parallel lines perpendicular to and directed towards the surface of the disc on which Greece is situated. Aristotle, on the other hand, holds that the earth is a large sphere, much larger than the surface area of Greece, and that it is situated at the centre of a universe organized in a series of concentric shells, whose radii directed toward the centre determine the direction of fall at each point. Now clearly the word 'fall' as used by each of them is, in a sense, loaded with his own theory. For Anaximenes it refers to a preferred direction uniform throughout space; for Aristotle it refers to radii meeting at the centre of the earth. But equally clearly, while they remain in Greece and converse on non-philosophical topics, they will use the word 'fall' without danger of mutual misunderstanding. For each of them the word will be correlated with the direction from his head to his feet when standing up, and with the direction from a calculable point of the heavens toward the Acropolis at Athens . . .

Now suppose Anaximenes and Aristotle agree on a crucial experiment. They are blindfolded and carried on a Persian carpet to the other side of the earth. That it is the other side might be agreed upon by them, for example, in terms of star positions – this would be part of the intersection of their two theories. They now prepare to let go of a stone they have brought with them. Anaximenes accepts that this will be a test of his theory of fall and predicts, 'This stone will fall.' Aristotle accepts that this will be a test of his theory of fall and predicts, 'This stone will fall.' Their Persian pilot performs the experiment. Aristotle is delighted and cries, 'It falls! My theory is confirmed.' Anaximenes is crestfallen and mutters, 'It rises; my theory must be wrong.'

(Hesse 1980:97f.)

The moral that Hesse rightly draws from this tale is that theory testing does not depend on everyone meaning the same thing by their terms, nor on the existence of a neutral observation language. Even theory-laden data can play a straightforward role in evaluation.

THE REGRESSION-OF-INTERESTS EXPLANATIONS

The argument I have been considering so far has been from the underdetermination of theories by data to the conclusion that the actual choices made by scientists cannot be rational, that their choices must be the result of social causes. I now want to deliver the ultimate refutation of this argument.[2]

Let us begin by assuming that underdetermination does indeed imply that theory choice is the result of social forces. But if we have underdetermination in general then we would have it here too. That is, a particular theory T may serve a scientist's interests, but more than one theory will do that. In fact, just as there are infinitely many different theories which do equal justice to any finite set of empirical data, so also are there infinitely many theories which will do equal justice to a scientist's interests. Phrenology, for instance, may have served well the interests of the Edinburgh middle classes in the early nineteenth century. This is what Shapin argues. But the underdetermination thesis claims that there were infinitely many different alternative theories which also would have served those interests equally well. Why is one of these singled out to serve those interests rather than any of the infinitely many others? Why was phrenology chosen by the Edinburgh middle classes?

What we have are theories T_1, T_2, T_3, ... underdetermined by the empirical data. We imagine that one of them, say T_2, is chosen because it serves some interest, say I_1. This is the point at which champions of social causation stop; they are satisfied that they have explained the choice of theories made by the historical agent when they have pointed out the relevant interest of that agent. However, the question now arises, 'Why T_2?' There are infinitely many theories T_2, T_2', T_2'', ... which do equal justice to the data *and* to the interest I_1, so why was one singled out over the others? An answer might be that T_2 was chosen over T_2', etc., because it served interest I_2. But there are infinitely many theories T_2, T_2^*, T_2^{**}, ... which equally serve interest I_2 (as well as serving I_1 and accounting for the data).

This regress comes to an end, but it is not a happy end for the sociologist. Eventually a final interest I_n is evoked to explain the choice of T_2. As above, however, there are still infinitely many theories T_2, T_2^+, T_2^{++}, ... which do equal justice to interest I_n (and to all of the other interests and to the data as well). So the final choice is left completely unexplained. Neither data nor interest determines a preference of T_2 over T_2^+, etc.

I am assuming there are only finitely many interests. If there are an infinity of them then the regress will simply go on for ever. Of course, this, too, leaves the choice unexplained.

There is a way for the sociologist to get out of this sticky regress.

Instead of saying there are infinitely many choices available at any point, deny this and maintain that only a very few alternatives are really and practically available. An interest may well uniquely determine a choice from the available rivals. But this, of course, is exactly what the rationalist wants (i.e., rationality consists in choosing from the available rivals). If it is admitted, the underdetermination argument will not even get off the ground.

Steven Yearley puts his finger on an important related problem. Barnes and others stress that the relation between specific interests and specific belief is itself a 'contingent and indeterminate' one. To this Yearley rightly objects, 'If alternative belief systems cannot be unequivocally related to differing societal interests but can only be connected through (indeterminate, revisable) interpretations, then differences in belief systems remain as mysterious as ever' (Yearley 1982:360).

It seems there is but one conclusion to draw. The underdetermination argument for social causation, so far one of the most persuasive and effective considerations in the sociologist's arsenal, is utterly destroyed.

LANGUAGE MATTERS

Among sociologists of knowledge, the members of the Edinburgh school have a special eye for language. It is here that the social plays a particularly subtle and deceptive role. Meaning is not at all what we intuitively think it to be; consequently, the whole notion of natural classification, which is parasitic upon it, is highly problematic. Barry Barnes, who champions finitism in language, asserts that 'alternative classifications are *conventions* between which neither "reality" nor "pure reason" can discriminate. Accepted systems of classification are *institutions* which are socially sustained' (Barnes 1981:312). This is a radical assertion. How does he get to it? Barnes's point of departure is a long discussion by Kuhn in his 'Second thoughts on paradigms' (1974), where he describes how an adult would teach a child the meanings of 'duck', 'goose', and 'swan'. This would be done by ostension, by pointing to examples. The child would be reinforced or corrected ('No, that's not a goose, it's a swan') until she had mastered the concept.

As Barnes notes, no two instances are exactly alike; similarity relations come into play. Because relative similarity is the basis, alternative classifications are possible.

Hence what the child learns is the preferred arrangement of some community, rather than something insisted upon by nature itself. Nature does not mind how we make clusters from the vast array of similarities and differences we are able to discern in it; all that is required of such clusters is that they constitute a tolerable basis for further usage. The clusters are conventions; the similarity relations which concepts stand for are conventions.

(Barnes 1982:24)

'But', it might be objected, 'surely we need not teach the meaning of swan by examples; we could instead define the term as, say, "a large, white-plumaged, orange-billed bird".' Barnes is ever ready:

Certainly, such statements could have assisted learning. But they could not have done so unless the terms they contained, 'white', 'plumaged', etc. had been informative. The child would have had to know already what these new terms meant, or how the terms were properly used. But there are no terms for which meaning or use is self-evident; nor does meaning accompany a concept as a mysterious halo. Hence any use of rules and definitions in conveying meanings . . . relies on earlier ostensive acts of learning . . .

(Barnes 1982:26f.)

And from this Barnes draws his grand conclusion:

It follows that *all* systems of empirical knowledge must rely upon learned similarity relations transmitted by ostension of practical demonstration, and that what any given term in such a system refers to can never be characterised without reference to learned similarity relations, i.e., to finite clusters of accepted instances of terms. Knowledge is conventional *through and through*.

(Barnes 1982:27)

So much for ducks; but what about the stuff of real science? Everything said about everyday things goes for theoretical entities as well: 'Paradigms are to concepts like "mass" and "force" what concrete instances are to concepts like "cat" and "shirt"' (Barnes 1982:52).

The implications of this are enormous. It means that at every step, every time we encounter a new object, it is a conventional decision to include it in the set of swans or to exclude it from that set.

concept application is a matter of judgement at the individual level, of agreement at the level of the community: it is open-ended and revisable. Nothing in the nature of things, or in the nature of langauage, or the nature of past usage, determines how we employ, or correctly employ, our terms.

(Barnes 1982:30)

There are no facts of the matter which we would be getting right or wrong, according to Barnes. If we started to call my car a duck, we would be changing a convention but we would not be mistaken. Of course, we would fall foul of the generalization that all ducks have webbed feet – but even that we could rectify, as long as we are prepared to accept that some webbed feet are manufactured by Goodyear and that it is a good idea to kick them when buying a used duck.

Barnes offers diagrams to show the differences between extensionalism and finitism, and to show the role of interests in the further development of concepts (see figures 3.1 and 3.2).

Kuhn is given credit for all this. I very much doubt whether Kuhn is a finitist himself, even if he does hold a conventionalist view of classification schemes. Barnes almost admits this, but remarks that it is not important since what really matters is that 'Kuhn's work . . . lends support to a finitist position, and helps us to see

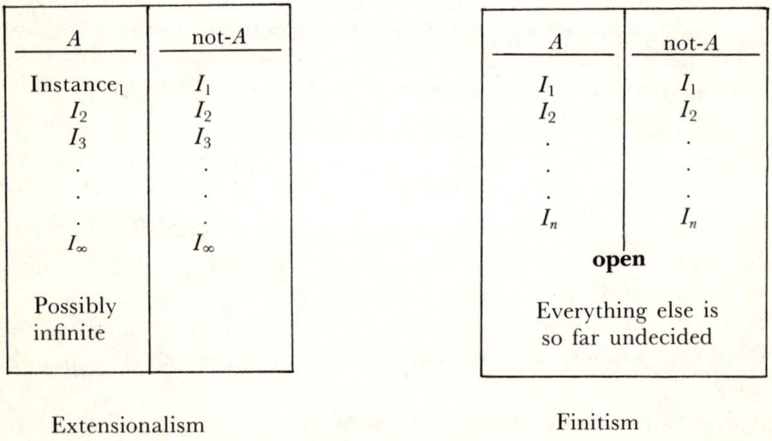

A	not-A
Instance$_1$	I_1
I_2	I_2
I_3	I_3
.	.
.	.
.	.
I_∞	I_∞
Possibly infinite	

Extensionalism

A	not-A
I_1	I_1
I_2	I_2
.	.
.	.
.	.
I_n	I_n
open	
Everything else is so far undecided	

Finitism

Source: Barnes 1982:31

Figure 3.1 The differences between extensionalism and finitism

Goals and Interests

Accepted act of
concept application

Existing similarity
relation (old
knowledge; usage at
earlier time)

Developed similarity
relation (new
knowledge; usage at
later time)

Source: Barnes 1982:112

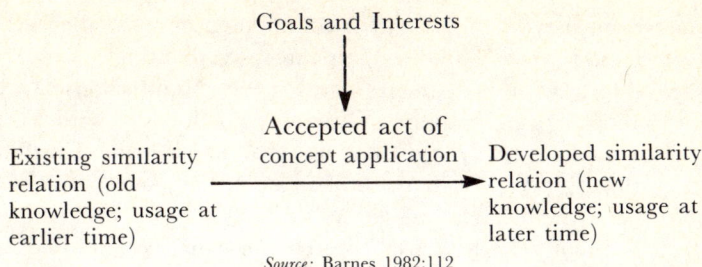

Figure 3.2   The role of interests in the further development of concepts

how knowledge can be understood in this sociologically interesting
way' (Barnes 1982:35f.).

Kuhn's duck–swan–goose example is by itself unconvincing. For
Kuhn the example was an illustration, not evidence. No one could
be won over to Barnes's conventionalism in language just from
considering it. There are better examples. One is provided by
Barnes himself in a different context (Barnes 1981). In Karam
taxonomy bats are classified with birds that can fly, while
cassowaries are˙ outside this classification. In our standard
classification scheme, the cassowary, though it is flightless, would
be included with other birds, and bats would be excluded. Barnes
describes some of the details involved in the Karam scheme, which
are often more subtle than the obvious distinction between 'able
to fly' and 'not able to fly'. In the next chapter we shall see the
example of the discovery (or invention) of the hormone TRF(H),
which has been well documented by Latour and Woolgar. Like
Barnes, they claim TRF(H) was not there waiting to be correctly
classified, but is, rather, a social construction. This example is
much more plausible than ducks and geese; so it is best to take
Barnes's claim seriously in spite of his unconvincing example.

Besides Kuhn, another important stimulus and precursor to
contemporary finitism is Wittgenstein (whose influence on Kuhn is
also well known). Wittgenstein held that *meaning is use*. 'What', he
asked, 'do all games have in common?' There would appear to be
ready counter-examples to anything we might propose: Played for
fun; follows rules; score is kept; has winner and losers. There are
lots of games which violate any one of these characterizations.
What then makes a game a game? According to Wittgenstein, it

would seem, they have nothing in common except that they are all called 'games'. We would not violate any defining rule or act contrary to any pre-existing essence, if we should choose to call some arbitrary thing a 'game'. Others might disagree with us, but no one would be really right or wrong. What is true of ordinary games, thought Wittgenstein, is also true of languages. Indeed, he came to call them 'language-games'.

David Bloor characterizes finitism in his recent book on Wittgenstein (which is perhaps better seen as an excellent introduction to the sociology of knowledge) as being

> the thesis that the established meaning of a word does not determine its future applications. The development of a language-game is not determined by its past verbal form. Meaning is created by acts of use. Like a town, it is constructed as we go along. . . . The label 'finitism' is appropriate because we are to think of meaning extending as far as, but no further than, the finite range of circumstances in which a word is used.
>
> (Bloor 1983:25)

And by way of providing a contrast, Bloor further remarks,

> At the heart of the rival view is the idea that predicates have associated with them a 'reference class' or 'extension'. The extension of a word is the class of all things of which it may be truly predicated. The extension of 'water' is everything, known and unknown, that could truly be called water.
>
> (Bloor 1983:25)

Before trying to criticize this account of meaning, let us first have a serious look at the alternatives. Though I think it is ultimately wrong, the 'finitist' account will take on some plausibility when we see the shortcomings of its 'extensionalist' rivals.

DESCRIPTIONS AND INCOMMENSURABILITY

A distinction is usually made between the *intension* of a term and its *extension*. ('Meaning' is often ambiguous between these two.) An intension (also, sense or connotation) is a unique description, a defining characteristic; it is what the dictionary is full of. To give the meaning of 'gold' as 'a precious yellow malleable ductile metal of high specific gravity' is to give the intension of the term. The extension (also, reference, denotation) is the physical stuff itself, all the actual gold that exists in the universe. The intension of 'leprechaun' is 'diminutive sprite who resides in Ireland' while the extension is the set of all leprechauns, the same extension as the set of all unicorns, namely, the empty set.

A description theorist holds that the key to meaning is the intension. Meaning is a perfectly accessible public entity and it is essential to our understanding of language. In contrast to the finitist view of meaning, this is quite properly thought of as an extensionalist account, since it views the situation as follows. Once the definition of a term is given, the set of objects which satisfy that definition is picked out; the intension determines the extension. We may not know whether the extension is empty or not, whether it has finitely many members or infinitely many; we may be mistaken about which objects are in it and which are not. But once determined by the defining description, the extension exists completely independently of us.

Within the description camp there are a variety of positions. Frege (1892) and Russell (1905) would tie a term to a unique description, but others such as Wittgenstein (1953) and Searle (1970) take the meaning of a term to be a cluster of descriptions. The latter view has its advantages, for it allows us to make discoveries that the former view does not. If Homer is defined as the author of the *Iliad* then it is impossible for scholars to discover that Homer didn't write the book. They could only conclude that there never was a Homer. But if Homer is defined by means of a cluster of descriptions: author of the *Iliad*, author of the *Odyssey*, the blind poet, then any one of these descriptions can be dropped or more new ones added. No element of the cluster is either necessary or sufficient for being Homer. So long as the cluster does not change too many of its elements the meaning of the term is unchanged. What goes for a proper name like 'Homer' also goes for natural-kind terms, the concepts of science, such as 'gold', 'water', 'energy', 'atom', etc.

Whether we stick with the single description of a term as providing the meaning or liberalize to a cluster account we are still faced with a serious dilemma. Meaning is intimately tied to belief. If we change one we modify the other. Scientific textbooks are not noted for their subtlety on these matters, but there are exceptions. Here is a fine expression of the linking of belief with meaning:

that view is out of date which used to say 'define your terms before you proceed'. All the laws and theories of physics . . . have this deep and subtle character, that they both define the concepts they use . . . and make statements about these concepts. Contrariwise, the absence of some body of theory, law, and principle deprives one of the means properly to define or even use concepts. Any forward step in human knowledge is truly creative in this sense: that theory, concept, law, and method of measurement – forever inseparable – are born into the world in union.

(Misner, Thorne, and Wheeler 1973:71)

What we mean by 'Homer', 'gold', 'energy' and 'atom' depends on what we believe about Homer, gold, energy and atoms. Perhaps our beliefs about Homer and gold have not changed much over the years, but this is not so for energy or atoms. Think first of the Greek beliefs about atoms, then Dalton's, and finally the contemporary quantum mechanical account of the nature of the atom. The belief change has been exceedingly radical. The various descriptions that would have been successively offered as defining the term 'atom' are at serious odds with one another. There is probably not a single description that could be said to be common to all three accounts. That means that even on a cluster account, there is no common element.

Consequently, the meaning of atom has changed. What the Greeks meant, what Dalton meant, and what contemporary chemists mean by 'atom' are three different things. The term is just a homonym. Should one say 'Atoms are thus and so', while another says 'Atoms are not thus and so', the contradiction is only apparent. This is the problem of incommensurability. Rival beliefs are not rival beliefs about the same thing.

Here is the difficulty: Can we have belief change without having meaning change? Can we change our minds about the nature of X and still be talking about X? There are two possible ways around the difficulty. One way is simply to deny that we ever do have radical belief change. Thus, we all mean the same thing simply because we all (more or less) believe the same thing. This is Donald Davidson's approach. The other approach is to hold that meaning has nothing to do with belief; instead, when we speak there is some sort of causal connection with what we are talking about. Thus, we can change our beliefs all we like, we will still be referring to the same thing. Kripke (1980) and Putnam (1975) introduced this as a theory of meaning and it has been picked up by others as a way to get around incommensurability. Let's look in turn at both of these attempts to save us from our difficulty.

DAVIDSON TO THE RESCUE?

A conceptual scheme is often characterized as a network of meanings and beliefs which is incommensurable with any other. Donald Davidson (1973) thinks the very idea of a conceptual scheme is empty.

Davidson starts with the assumption that there is uniquely associated with a conceptual scheme a single language, or rather a set of inter-translatable languages. For example, associated with the conceptual scheme of classical mechanics is the language *Newtonian English* and also the language *Newtonian French*. We can express the same thing in either language. In English:

Force = mass times acceleration.

and in French:

Force = masse multiplié par accélération.

Davidson then understands the claim of the existence of distinct conceptual schemes as being the claim that the associated languages are not inter-translatable. Letting CS and CS' be two conceptual schemes which are associated with the languages L and L', respectively, we have:

$CS' \neq CS'$ if and only if L and L' are *not* inter-translatable.

Davidson ultimately contends that 'nothing . . . could count as evidence that some form of activity could not be interpreted in our language that was not at the same time evidence that that form of activity was not speech behaviour' (Davidson 1973:7). In other words, the very idea of a conceptual scheme is incoherent. He proceeds to this conclusion by first considering the relation between the *adequacy* of a conceptual scheme, on the one hand, and its *truth*, on the other. They are, he contends, the same:

The trouble is that the notion of fitting the totality of experience, like the notions of fitting the facts, or being true to the facts, adds nothing intelligible to the simple concept of being true. To speak of sensory experience rather than the evidence, or just the facts, expresses a view about the source or nature of evidence, but it does not add a new entity to the universe against which to test conceptual schemes. The totality of sensory evidence is what we want provided it is all the evidence there is; and all the evidence there is is just what it takes to make our sentences or theories true. Nothing, however, no *thing*, makes sentences and theories true: not experience, not surface irritations, not the world, can make a sentence true. That experience takes a certain course, that our skin is warmed or punctured, that the universe is finite, these facts, if we like to talk that way, make sentences and theories true. But this point is put better without mention of facts. The sentence 'My skin is warm' is true if and only if my skin is warm. Here there is no reference to a fact, a world, an experience, or a piece of evidence.

(Davidson 1973:16)

From the claim that an adequate conceptual scheme is one which is true, Davidson passes to the view that the difference between some other scheme and ours is that both are true (or largely true) but not translatable:

Our attempt to characterize languages or conceptual schemes in terms of the notion of fitting some entity has come down, then, to the simple thought that something is an acceptable conceptual scheme or theory if it is true. Perhaps we better say *largely* true in order to allow sharers of a scheme to differ on details. And the criterion of a conceptual scheme different from our own now becomes: largely true but not translatable.

(Davidson 1973:16)

There is a simple mistake here; Davidson has passed from stating a criterion for a conceptual scheme being (largely) true to saying implicitly that *ours* is (largely) true. At most Davidson could make his claim for two conceptual schemes which were totally adequate to the evidence, or as he would have it, two schemes which were both largely true. But it is absurd to make such a claim for two arbitrary schemes. One framework, or even both, may be largely inadequate. Certainly those who think there are such entities exemplified in the history of science do not think that all of them were adequate/true. Quite the contrary. Kuhnian paradigms do, after all, have crisis periods. Feyerabend (1975) thinks that every theory is in some sort of serious trouble. Lakatos's (1970) research programmes and Laudan's (1977) research traditions are more or less progressive; they are certainly not adequate (or even largely adequate), otherwise there would be no point to progressing. The problem of scientific rationality, which is what many champions of conceptual schemes are grappling with, is the problem of choosing between rival *inadequate* conceptual schemes. Since Davidson's argument rests on the false assumption that rival schemes will be adequate, it is unsound. But there is more to his argument, so I will push on to the next step.

The claim '"Snow is white" is true if and only if snow is white' strikes us as trivial. And so it is; but it is also an ideal criterion for a theory of truth. An adequate theory of truth must capture all statements of that form. This is Tarski's criterion (not his theory of truth) and it is known as:

CONVENTION T: a satisfactory theory of truth for a language L must ential, for every sentence *s* of L, a theorem of the form '*s* is true if and only if *p*' where '*s*' is replaced by a description of *s* and '*p*' by s itself if L is English, and by a translation of *s* into English if L is not English.

(Davidson 1973:17)

Note that Convention *T* employs the notion of a translation; something can be true only if there is an appropriate translation. Davidson uses this to justify the claim that there can be no notion of truth at all which is not tied to the idea of translation. We will not be able to say something is true unless we can translate it into our conceptual scheme, that is, into our language.

Since Convention *T* embodies our best intuition as to how the concept of truth is used, there does not seem to be much hope for a test that a conceptual scheme is radically different from ours if that test depends on the assumption that we can divorce the notion of truth from that of translation.

(Davidson 1973:17)

Davidson's strategy should now be evident. His argument runs:

1. Suppose *CS* and *CS'* are two different conceptual schemes.

2. Thus, their associated languages L and L' are not inter-translatable.
3. Assume CS and CS' are both totally adequate to the evidence.
4. Thus, they are both true (or largely true).
5. By the best account of truth we have (i.e. Convention T), L and L' are both translatable into English.
6. Hence, L and L' are inter-translatable.
7. Therefore, contrary to the initial supposition, CS and CS' are not different conceptual schemes.

So it is established that translation is indeed possible, but how, as a practical matter, is it carried out, especially given the intimate connection between meaning and belief?

> If all we know is what sentences a speaker holds true, and we cannot assume that his language is our own, then we cannot take even a first step towards interpretation without knowing or assuming a great deal about the speaker's beliefs. Since knowledge of beliefs comes only with the ability to interpret words, the only possibility at the start is to assume general agreement on beliefs. We get a first approximation to a finished theory by assigning to sentences of a speaker conditions of truth that actually obtain (in our own opinion) just when the speaker holds those sentences true. . . . Charity is forced on us; – whether we like it or not, if we want to undertand others, we must count them right in most matters.
>
> (Davidson 1973:18f.)

So, to sum up Davidson's position, the idea of incommensurable conceptual schemes is an incoherent one, and furthermore, not only can we understand one another, but, by the principle of charity, we all largely believe the same things.

As I pointed out above, assumption 3 certainly will not stand as representing the real state of affairs. Proponents of conceptual schemes take the actual schemes to be transitional devices. They are not claimed to be adequate at all; the history of science, including contemporary science, is a history of *inadequate* schemes. (Though some have been better than others, rationality consists in choosing the most adequate from among the available rivals – the completely adequate is simply not available.)

This rather simple consideration upsets Davidson's argument. Since the real history of science does not instance completely adequate conceptual schemes, Davidson's considerations, even if otherwise cogent, do not apply. A state of non-translatability may well obtain between conceptual schemes which are both inadequate.

The problem with the Davidsonian argument is similar to the problem with many underdetermination arguments. They are both based on the ideal case; the latter on having infinitely many different theories all equally compatible with the data; the former on having conceptual schemes which are totally adequate. There is nothing in principle wrong with idealizations. In a physics problem, one could attribute zero mass to a pulley or zero friction to an inclined plane and still gain considerable insight into a given physical set-up. But insight would not be gained if one attributed zero mass to a test particle in a gravitational field in the hope of better understanding the nature of gravitational force. In this case as well as in the underdetermination and Davidsonian cases, the idealization perverts an essential feature of any real case. Real theories are inadequate.

But let us end on a positive note, even though the brief considerations to follow have no bearing on the general argument so far. There may be something of interest left in Davidson's view, so let's not toss it out. Suppose we maintain the idea that some schemes are more adequate than others, and in particular, that some schemes could be totally adequate. Then what Davidson's argument might be said to demonstrate is that 'in the end of science' when enquiry is 'finished' the notion of a conceptual scheme will be empty. There won't be two distinct, totally adequate frameworks. Accordingly, we might say that conceptual schemes exist only in *inadequate* forms, and thus should be seen as epistemological instruments which can be tossed away like Wittgenstein's ladder when their work is done. But we can't toss them away just yet.

SAVED BY THE CAUSAL THEORY?

Davidson hasn't given us a plausible theory of language, but perhaps the causal theorists can. We turn our attention now to a new and quite popular account of meaning.

Hilary Putnam (1975) notes that traditional theories of meaning, such as the descriptional account above, hold two things as central:

1. To know the meaning of a word is just to be in a certain psychological state.
2. The meaning of a word determines its reference; that is, two words with the same intension must have the same extension.

Putnam denies both of these. (Finitists, too, would deny the second, but their reasons for denial would be quite different.)

Putnam presents us with an intriguing philosophical thought-experiment which is designed to convince us of the wrong-headedness of description accounts of meaning and to lead us rather naturally to Putnam's own theory. Imagine a world, Twin Earth, just like ours down to the last noticeable detail. There is a city there called 'Toronto' which is on 'Lake Ontario', and it's full of stuff the people there call 'water'. On Earth 'water' means H_2O, but on Twin Earth the stuff in their Lake Ontario has a different chemical composition, say, XYZ. What the Twin Earthers mean by 'water' then is not H_2O, but rather XYZ. If our chemists went to Twin Earth, says Putnam, they would report something like this: 'There is a substance there in the local lakes that looks like water, tastes like water, is used by the locals just as water is used on Earth, and they even call it "water"; but it isn't water, it is XYZ.' Not only would our scientists say this, Putnam claims, but it is exactly what they should say.

We might even imagine that on Twin Earth each of us has an exact duplicate, a *Doppelgänger*. Suppose I am in the right psychological state so that on the traditional theory of meaning I know the meaning of 'water'. My Twin Earth counterpart is in exactly the same psychological state. However, we mean different things; I mean H_2O while he means XYZ. Thus, contrary to descriptional and other traditional theories of meaning, being in a specific psychological state does not determine a unique meaning. Moreover, the intension of 'water' on Earth and the intension of 'water' on Twin Earth are exactly the same; yet they mean quite different things. Consequently, intension does not determine the extension.

Though intensions go by the board, we nevertheless possess a stereotype of water. It is the same stereotype as the Twin Earthers have for the stuff they call 'water', namely, 'a clear liquid, etc.' This stereotype embodies beliefs about water, just as intensions did, but it has nothing to do with meaning.

If having the intension won't do the trick, how then do we manage to refer? How do we manage to be talking about water, wallabies, or w-particles when we say 'water', 'wallabies', or 'w-particles'? Putnam and Kripke have a story to tell; it is a different story in each case, but the general picture is like this. There is an initial baptism. Someone says of the stuff in Lake Ontario, 'This is water.' Or someone in ancient Egypt says of a new baby, 'Moses'. Then through a causal chain linking us to the baptism of water or of Moses, we are able to talk about these things, we are able to refer to them.

Of course, we'll never know how initial baptisms of water went, but let us assume for the sake of our story that our ancestors stood around Lake Ontario and said 'Water'. Similarly, the Twin Earthers stood around their Lake Ontario and also said 'Water'. Later we discovered that water is H_2O. This discovery is a piece of fallible empirical science; perhaps water isn't H_2O after all. But if we're right, then *necessarily* water is H_2O. In any possible world in which there is water, it must have this micro-structure. Consequently, the stuff the Twin Earthers called 'water' simply isn't water.

For our purposes here, the crucial thing is that what 'water', or any other natural-kind term, refers to has nothing to do with our beliefs about it. Intensions, defining characteristics, according to Putnam and Kripke, do not detemine the reference. This suggests, in consequence, that it might be possible to change our beliefs without changing meanings, that is, without changing the reference. The link between theory and meaning would be broken and the problem of incommensurability overcome. Among others, William Newton-Smith and Ian Hacking have quite explicitly thought the causal theory will save the day.

It was once the accepted wisdom that a word such as 'electron' gets its meaning from its place in a network of sentences that state theoretical laws. Hence arose the infamous problems of incommensurability and theory change. For if a theory is modified, how could a word such as 'electron' go on meaning the same? . . . Putnam saved us from such questions by inventing a referential model of meaning. . . . Serious discussion of inferred entities need no longer lock us into pseudo-problems of incommensurability and theory change. Twenty-five years ago the experimenter who believed that electrons exist, without giving credence to any set of laws about electrons, would have been dismissed as philosophically incoherent. Now we realize it was the philosophy that was wrong, not the experimenter.

(Hacking 1984:158ff.)

We have been referring to the same thing all along; it is only the stereotype of electron that has changed, not the meaning.

Newton-Smith has embraced a modified version of the Putnam account. Since he gives details of how it is to work, his version is worth looking into. The idea is that we do not *define* a term by means of its believed properties, nor even by means of its causal effects. Rather we *refer* to it through its causal effects.

x is that magnitude which is causally responsible for certain specified effects.

(Newton-Smith 1981:171)

Different scientists might introduce a term via different effects. Thus,

Scientist$_1$:   Term$_1$ is that which is responsible for Phenomenon$_1$.
Scientist$_2$:   Term$_2$ is that which is responsible for Phenomenon$_2$.

If Scientist$_2$, who we may suppose arrives on the scene at a later date, says 'Term$_2$ is responsible for Phenomenon$_1$' then we may conclude that Term$_1$ = Term$_2$.

The view seems quite plausible. We might imagine J. J. Thompson introducing 'electron' in connection with photographic-plate phenomena while later scientists use the term in connection with cloud-chamber phenomena. Since the later scientists would also say that electrons are responsible for the markings on the photographic plates, it seems entirely reasonable to conclude that what they mean by 'electron' and what Thompson meant are the same. Thus, competing scientists may have rival things to say about electrons, but they are talking about the same thing. The meaning of a term is quite independent of theories about it.

Appealing though it is, there are a couple of problems with Newton-Smith's account. For one thing he takes 'Phenomenon$_1$' to be something describable in both theories. It may well be that the second scientist would not even recognize such a thing. Given that we are in the middle of a discussion of incommensurability, the existence of a neutral description of phenomena is one thing that cannot be taken for granted.

Second, suppose we change the illustration from electron to phlogiston. Let the two scientists be Priestley and Dalton, and let the terms introduced by each be 'Phlogiston' and 'Oxygen'. Suppose Dalton later says 'Oxygen is responsible for Phenomenon$_1$ [as well as for Phenomenon$_2$]'. We must then conclude that phlogiston = oxygen, which it surely isn't.

These problems are with the attempt to use a causal theory of meaning to get round the problem of incommensurability. But there are more general difficulties with the Putnam–Kripke theory apart from this particular application. D. H. Mellor (1977) notes that often in the past we have discovered that a substance and its isotope are both named, say, 'chlorine'. What we have discovered is that there are two types of chlorine. The moral of such historical examples is that should we ever get to Putnam's Twin Earth we would say that we have discovered another type of water. This reaction, says Mellor, is the correct one, and quite of a piece with traditional theories of meaning.

There is yet another kind of basic objection I want to add. According to Putnam, when we baptize water or any other natural kind we pick out a bit of it and intend our term to apply to all the stuff in the universe that is the same. As Putnam puts it, x is water if and only if x bears the relation *same L* (for same liquid) to what we are here referring to as 'this'. Of course, we might accidentally fail to realize our intentions; we might call a glass of gin 'water' thinking we were baptizing something taken from Lake Ontario. Since our intentions do matter (when it comes to picking things out initially), this sort of mishap is no problem; we are not bound to call gin 'water' for evermore just because of this initial mistake.

The discovery that some x does indeed bear the relation *same L* to water is a bit of fallible science. We may be wrong that water is H_2O, or that the stuff on Twin Earth is XYZ. Putnam's claim is only that if we are right then that is the way water *must* be, and so, the stuff on Twin Earth just isn't water. It is clear from his discussion and examples that *same L* means 'same micro-structure'. However, the discovery that something is or is not the *same L* as water is doubly fallible.

In our present mechanistic view of how things work, micro-structure plays an all-important role. Causality is efficient causality. But this is not the only way to see the world (though I think it is the correct way). Aristotle's outlook was teleological; what mattered was purpose, not composition. Explanations were in terms of functions or purposes, not micro-structures. Suppose a teleological account of the world is the right one. Then *same L* would mean something like 'has the same purpose', 'plays the same role', or 'performs the same function'. If this were the accepted theory then, our scientists visiting Twin Earth would say, the stuff Twin Earthers call water is made of XYZ instead of H_2O, but it falls from the sky, nourishes organisms, puts out fires, and so on, therefore it performs the same function as on Earth, so it must be water.

This in itself may not be a problem. Putnam could reply that, yes, *same L* is doubly fallible in its determination, but this does not imply that the theory is wrong. At worst Putnam was merely hasty in assuming that micro-structure determined all. Perhaps the causal theory of meaning itself is untouched by this sort of example; however, its application to incommensurability is rendered hopeless.

Commensurability must surely be a symmetrical relation. If Earth scientists can truly say or truly deny that what Twin Earthers call 'water' is the same as what we call 'water', then Twin Earth scientists should be able to say the same. However, it is not necessary that their scientists hold the same theory as ours. Suppose, unlike us, the Twin Earthers hold a teleological account of nature as outlined above. Then they would say that our term 'water' refers to the same thing as their term since both fill the same functional role. However, our scientists say that what Twin Earthers call 'water' isn't water since it does not have the same micro-structure.

The problem of incommensurability has always been a philosophical issue. Now it turns out that whether there is or is not sameness of meaning depends on which scientific theory one adopts. Rather than solve the problem of the interconnection of belief and meaning, the causal theory of Putnam and Kripke has only added to the difficulties.

I have given over much space in the last few sections to discussing some of the issues in the philosophy of language. The topic, it seems to me, is completely up in the air. There is no theory available that one could point to with satisfaction and say to sociologists that they were simply throwing dust in our eyes. There are real problems everywhere. So new views with radical sociological implications deserve serious consideration, since they might be presenting us with the only way out of our conundrum. On the other hand, finitist approaches have their own short-comings. We get a glimmer of what they might be when we see Barnes standing back and looking at the philosophical fight from a sociological viewpoint.

BARNES ON THE DEBATE

Barnes thinks that champions of the causal theory and champions of the descriptions account are alike wrong in their views of how language works. He gives us his finitist alternative. But he has something more to offer; he has an account of the nature of the debate between the two extensionalist schools. Both sides in this philosophical debate may think they are trying to describe semantic reality, but in fact, according to Barnes, they are offering rival conventions concerning future usage:

> One might say that a theory of extension is not a theory related to actual accepted usage at all, but is an expression of philosophical *preferences*. If this is the case, then extensions are not indications of how accepted usage will actually proceed and develop, but indications of how philosophers happen to think it *ought* to develop: 'theories of extension' must then be seen as 'normative' rather than 'empirical' theories, as vehicles with which, as it were, philosophers moralize about language. . . .
>
> As far as actual usage is concerned, we have here a situation where choice is possible between two alternative modes of development. Which development is actually taken up and institutionalized can be no more than a matter of contingent collective judgment. Such a judgment must be understood in terms of contingent features of the setting wherein it is made, and cannot be understood simply by reference to the pre-existing semantic properties of the term involved. . . .
>
> Descriptivism and essentialism represent two strategies of development, and if one were to be adopted rather than the other, that would be nothing to do with any alleged inherent properties of terms. It would merely be that the application of

authority had extended custom in one direction rather than another.

(Barnes 1982a:30ff.)

It's an intriguing and provocative speculation. But could his sociological account of a philosophical debate even hope to be right? I think not. Of course, this is a peripheral issue for Barnes, who is merely giving an explanation of the existence of rival extensionalist theories where both sides can be taken to be equally wrongheaded. Should his account fail in its aim it would be of no great significance, since his finitism is not itself at issue here. Barnes is merely trying to explain the infighting between his rivals, and his theory of that battle is quite independent of finitism itself. At least, that is the way he would like to see things. It will turn out, however, that failure here is very instructive. It will indeed have implications for his own finitist theory.

The problem is a rather simple one. In order for Barnes to have something to explain, there must be a connection between the views of the description theorists and the causalists and the different linguistic usages which they, respectively, recommend. According to Barnes's finitism, however, there can be *no* such correlation. He has destroyed his own explanatory strategy.

To see this let me construct an analogy. Suppose we have rival political parties, *A* and *B*. Normally we understand parties as being vehicles for achieving various social ends, with different parties generally seeking different goals. When individuals *X* and *Y* are members of *A* and *B*, respectively, we can predict their political behaviour. That is, when some social issue arises we can predict (fallibly, but with reasonable success) which side of the issue *X* and *Y* will be on. Suppose, however, that no matter what *X* and *Y* do, their actions are compatible with being in either the *A* or the *B* party. Should this turn out to be the case then there is simply no point at all in saying that the parties *A* and *B* have rival or even any particular social aims. Yet this is exactly what Barnes would have us believe.

Barnes would have us think that descriptionists and causalists are not arguing over the way things are, but are instead advocating rival courses of future action. The descriptionist party policy is that water is a colourless, tasteless, liquid convertible by heat into ice or steam. The causalist party policy is that water is H_2O. Suppose *X* advocates the descriptionist policy and Y the causalist policy. Can we predict their behaviour when they newly encounter some particular substance? Yes, but only on one condition. If both agree that it is, say, a colourless, tasteless liquid, etc., and they agree that it is not H_2O, then we can indeed predict their behaviour: X will advocate calling it 'water', while Y will reject this policy.

But the condition required for making this prediction is denied us. We cannot, according to Barnes, determine that the new stuff 'really is', a colourless, tasteless liquid, or that it 'really isn't' H_2O. Finitism disallows the existence of party policies that can be applied to new cases in exactly the same way that it denies the existence of semantic rules that determine the extension of a concept.

If descriptionists went to Putnam's Twin Earth, then, according to Barnes, they would conventionally decide that the stuff there was or was not water because they conventionally decided whether it was or wasn't a clear liquid, etc. Causal theorists who go to Twin Earth will also conventionally decide that the stuff there is or isn't water because they will conventionally decide whether it is or is not H_2O. The two camps will differ in their accounts of what they have done, but which view of language they hold does not determine which convention either will adopt.

If Barnes sticks to his finitism, then the behaviour of the champions of either the descriptions account or the causal account is completely unpredictable. No matter what either does, it is completely compatible with the linguistic policy each advocates. On the other hand, if we want to hold that their behaviour is explainable and predictable, and Barnes clearly does, then finitism goes by the boards.

I wish I had a strong direct argument to show that language is not deeply conventional in the way it hooks onto the world. Instead, all I can conclude is that any sociological account of science needs a non-conventional view of meaning as well. But as consolation prizes go, that's not too bad.

Chapter Four

THE EXPERIMENTER'S SOCIAL CIRCLE

'Whereas we have fairly detailed knowledge of the myths and circumcision rituals of exotic tribes, we remain relatively ignorant of the details of equivalent activity among tribes of scientists . . .' So wrote Bruno Latour and Steve Woolgar in *Laboratory Life* (1979:17), which recounts their activities as 'anthropologists in the lab'. *Laboratory Life* is a modern classic; of more recent vintage is Harry Collins's *Changing Order* (1985). Both are part of the rapidly growing literature on the nature of science, especially experimental science, written from a sociological or anthropological perspective.

Studies like those by Collins and by Latour and Woolgar have several advantages over sociological studies of great historical events. Not only are they the result of firsthand experience of daily scientific life, but the examples tend to be more mundane. Collins does three case studies on the replication of a laser, the detection of gravity waves, and the detection of paranormal phenomena; Latour and Woolgar report on the isolation and study of TRF(H), a substance produced by the edocrine system. The subjects of study are not the glamorous Newtons, Darwins, Pasteurs, or makers of modern quantum mechanics. Rather they are the everyday scientists with no reputations outside their own scientific speciality, possibly no reputation at all. These are not epic struggles which will greatly change the course of history; but what they lack in glamour they make up for in being typical.

There are exceptions, though. The work studied by Latour and Woolgar won a Nobel Prize (Guillemin and Schally, medicine, 1977). Good for the scientists; too bad for Latour and Woolgar. But at least they studied great science in the making, that is, before it was publicly certified as great science.

GRAPHOMANIA

With galling frankness, Latour and Woolgar set the scene for their anthropological investigation:

> We take the apparent superiority of the members of our laboratory in technical matters to be insignificant . . . This is similar to an anthropologist's refusal to bow before the knowledge of a primitive sorcerer . . . There are . . . no a priori reasons for supposing that the scientist's practice is any more rational than that of outsiders.
>
> (Latour and Woolgar 1979:29f.)

The upshot is to deflate any pretensions scientists may have about their own activities and to undermine the common starting-point for rationalist analyses of science. Whether scientists are good at getting at the truth, or if unearthing facts is even what they are trying to do, is something which will have to be established, not assumed.

Of course, it is quite fitting to minimize one's assumptions, and, in particular, to not assume a priori that scientists must be rational seekers after the truth. But do Latour and Woolgar conclude as a result of their anthropological investigation that this is nevertheless so? They do not. Amazingly, Latour and Woolgar interpret scientists as a society of graphomaniacs. They 'make sense of the laboratory in terms of a tribe of readers and writers who spend two-thirds of their time working with large inscription devices' (1979:69). Continuing, Latour and Woolgar remark that scientists

> appear to have developed considerable skills in setting up devices which can pin down elusive figures, traces, or inscriptions in their craftwork, and in the art of persuasion. The latter skill enables them to convince others that what they do is important, that what they say is true, and that their proposals are worth funding. They are so skillful, indeed, that they manage to convince others not that they are being convinced but that they are simply following a consistent line of interpretation of the available evidence.
>
> (Latour and Woolgar 1979:69f.)

Publication is not the means whereby science achieves its goals; publication *is* the very goal of science. All that mucking about in the lab has no other purpose than a literary one. Experiments done, calculations made, consultations had are all for the lab's output – words, words, words. A prosaic analysis of science if ever there was one. But isn't this just 'publish or perish' gone mad? Latour and Woolgar don't think so. Writing of themselves in the third person, they claim to be very capable of resisting the sensible:

> the anthropologist feels vindicated in having retained his anthropological perspective in the face of the beguiling charms of his informants: they claimed merely to be scientists discovering facts; he doggedly argued that they were writers and readers in the business of being convinced and convincing others. Initially this had seemed a moot or even absurd standpoint, but now it appeared far more reasonable. The problem for participants was to persuade readers of papers (and constituent diagrams and figures) that its statements should be accepted as fact. To this end rats had been bled and beheaded, frogs had been flayed, chemicals consumed, time spent, careers had been made and broken, and inscription devices had been manufactured and accumulated within the laboratory. This, indeed, was the very raison d'être of the laboratory. By remaining steadfastly obstinate, our anthropologist observer resisted the temptation to be convinced by the facts. Instead, he was able to portray laboratory activity as the organization of persuasion through literary inscription.
>
> (Latour and Woolgar 1979:88)

Such powers of resistance are only to be marvelled at.

MOTIVATION AND CREDIT

The subtitle of Latour and Woolgar's book is 'The Social Construction of Scientific Facts'. The main thesis is just what the subtitle suggests: facts are a social construction, the result of various negotiations of a political character among the interested parties. Before looking in detail at this, a second thesis which is just as important will be briefly examined.

Why do scientists do science? Or, as Latour and Woolgar put it,

> What drives scientists to set up inscription devices, write papers, construct objects, and occupy different positions? What makes a

scientist migrate from one subject to another, from one laboratory to another, to choose this or that method, this or that piece of data, this or that stylistic form, this or that analogical path?

(Latour and Woolgar 1979:189)

Their answer is unexceptional and entirely plausible: peer recognition. Scientists want the approval of other members of their tribe. Many scientists are quoted; one for example, says, 'I was a physician . . . but I wanted positive feedback proving my smartness . . . patients are not so good for that . . . I wanted a very rare commodity; recognition from peers . . . [so] I moved to science' (1979:190). Scientists work hard so that they will get credit for it; credit and future credibility are their rewards. In a qualified way Latour and Woolgar make much of this:

it would be wrong to regard the receipt of reward as the ultimate objective of scientific activity. In fact, the receipt of reward is just one small portion of a large cycle of credibility investment. The essential feature of this cycle is the gain of credibility which enables reinvestment and the further gain of credibility. Consequently, there is no ultimate objective to scientific investment other than the continual redeployment of accumulated resources. It is in this sense that we liken scientists' credibility to a cycle of capital investment.

(Latour and Woolgar 1979:198)

But do successful capitalists only make money? Don't they also make carpets and clothes pegs? And, in fact, don't they make money because they make artefacts? Undoubtedly some scientists get more credit than they deserve, just as capitalists make more money than they deserve. But if we accept the economic analogy that Latour and Woolgar propose, it is hard to resist the conclusion that scientists get credit because they do (good) science.

I wouldn't for a moment deny the existence of credit in the workings of science; I want only to assert that it seems perfectly obvious that credit is the result of something else. Nor do I hold that past credit will have no impact on future science. Yet Latour and Woolgar suggest that any rationalist would indeed deny this. They remark that 'From the perspective of some epistemologists, we would expect the reliability of data to be an issue quite

distinctly separated from the evaluation of individuals in the field'
(1979:200). To show this is typically not so, they cite one of their
recorded conversations. Here two scientists are doing an experiment;
one is giving a running commentary both on the experiment and
on his confidence in a third scientist:

> I bet you the peptide is going to do nothing . . . this is the
> confidence I have in my friend T. [C squeezed the syringe and
> enjoined the rat]: O.K., Charles T., tell us. [A few minutes
> passed.] See, nothing happened . . . if anything the rat is even
> stiffer [sigh]. Ah, my friend T . . . I went to his laboratory in
> New York and saw his records . . . which led to
> publication . . . it made me feel uncomfortable.
>
> (Latour and Woolgar 1979:202)

The case study that Latour and Woolgar have undertaken is
supposed to 'provide a telling argument for the feasibility of the
strong programme in the sociology of science' (1979:106). So far it
has done nothing of the sort. The essence of the strong programme
is that it perceives social factors right in the very content of scientific
theories. The features that Latour and Woolgar point to above
don't do this at all.

The fact that scientists evaluate data and their colleagues
simultaneously says nothing about the content of any given theory.
Suppose I have grounds for thinking that the chip in my calculator
is malfunctioning (every now and then it adds two and two and
gets five), and that I take this into account when forming my
beliefs. (That is, my beliefs are based on the results of my
computer-processed data as well as my lack of faith in the
computer itself.) Is this a sociological response? Do my subsequent
beliefs have sociological content?

We evaluate individual scientists on the basis of their track
records. In the future we will be suspicious of those scientists who
have produced data thought to be faulty in the past, just as I am
suspicious of my pocket calculator. Obviously, science is a group
enterprise; no scientist does all the interacting with nature in order
to accumulate the necessary data. Tables of integrals are taken for
granted, and so are the contents of *The Handbook of Chemistry*. The
results of the latest experiments meet a wide spectrum of attitudes.
Most often differences in attitude are due to the nature of the
experiment itself; if it involves straightforward techniques, the

result is relatively unproblematic. If it involves apparatus which is thought questionable (for whatever theoretical reasons), then the results are thrown in doubt. Scepticism surrounded Weber's data on gravitational waves, not because he was thought by all his colleagues to be an incompetent experimenter, but because of the extreme sensitivity of his equipment. It is much more rare to make a judgement of outright incompetence of a particular experimenter. But when it happens, is it any different from the pocket calculator case? Of course not. In none of these cases is a sociological analysis called for.

CONSTRUCTING FACTS

One of the many conversations Latour and Woolgar cite goes as follows. It is intended to illustrate what they call 'the microprocessing of facts'.

> Wilson: Anyway, the question for this paper is what I said in one of the versions that there was *no evidence* that there was any psychobehavioural effect of these peptides injected I.V. . . . Can we write that down?
>
> Flower: That's a *practical* question . . . what *do we accept* as a negative answer? [Flower mentioned a paper which reported the use of an 'enormous' amount of peptides with a positive result.]
>
> Wilson: That *much*?
>
> Flower: Yes, so it depends on the peptides . . . but it is very important to do . . .
>
> Wilson: I will give you the peptides, yes we have to do it . . . but I'd like to read the paper . . .
>
> Flower: You know it's the one where . . .
>
> Wilson: Oh, I have it, OK.
>
> Flower: The threshold is 1 μg. . . . OK, if we want to inject 100 rats (we need at least a few micrograms) . . . it's a practical issue.

> (Latour and Woolgar 1979:156)

This is followed by a rather long analysis. First, they point out that Wilson and Flower (fictitious names) are academic peers, and that they are writing a paper together. (Remember how important that is.) Do they show any sign of being led by the evidence? Not at all,

'Rather Flower's comment shows that it depends on *what they choose to accept as* negative evidence.' Each is expert in a different field; Wilson is a newcomer to the area they are working in at present. As such, Wilson probably needs the expertise of Flower for their joint paper more than Flower needs him. But Wilson has something to give: peptides. He has control over a rare substance, but he will give some to Flower, who needs it to make the paper look credible. But is this not just a clear case of marshalling the relevant evidence? It would seem not:

> When, for example, Flower says, 'it is very important to do . . . ' it is possible to envision a range of alternative responses about the relative importance of the uses of peptides. In fact, Wilson's reply ('I will give you the peptides') indicates that Wilson hears Flower's utterance as a request for peptides. Instead of simply asking for them. Flower casts his request in terms of the importance of the investigation. In other words, epistemological or evaluative formulations of scientific activity are being made to do the work of social negotiation.
>
> (Latour and Woolgar 1979:157)

As well as these negotiations between themselves, Flower and Wilson are also trying to establish what it will take, not to get the facts right, but to convince other people. And on and on the story goes. This is how facts, according to Latour and Woolgar, are constructed. No fact is ever discovered; they are all made.

The embarrassing thing about facts is the etymology of the word. It comes from the Latin *facere*, which means to make or construct. Anti-realists revel in this bit of lexicography. The history of the word, they suggest, reveals its true nature: facts are not already there waiting to be discovered; rather, they are made by us. But, of course, words change their meaning, and if the etymology of the word 'fact' is the only embarrassing thing about it, then realists and rationalists have little with which to be concerned.

We might put the problem we are now facing this way:

Are there things justly called 'facts' that scientists discover?
Are there things justly called 'facts' that scientists create?

These two questions seem to embody the rationalist/sociologist dichotomy: 'Facts are discovered' versus 'Facts are created'. But it is not a nice clean division of the options. The problem is simply this. Almost all contemporary rationalists are fallibilists; belief based exclusively on good reasons might still be wrong. All the evidence might have pointed to the correctness of the phlogiston theory, for example, even though (we now presume) there is no phlogiston. In such a case the 'facts about phlogiston' were certainly not discovered; they were in some sense, I suppose, 'made'.

The way to sort this out is just to make the obvious distinction between fact and what is believed to be a fact. Our beliefs are true when what is believed to be a fact is indeed a fact. It would seem to be an obvious distinction, hardly worth making. But Latour and Woolgar's study of TRF(H) rests on ignoring it.

Once the distinction is made we can ask distinct questions:

Did the scientists in Latour and Woolgar's study *create* the TRF(H) facts?
Did they *discover* the TRF(H) facts?
Did they have good reason to accept *what are believed to be* the TRF (H) facts?

Let's see why Latour and Woolgar think TRF(H) facts are human constructs.

TRF(H) (for Thyrotropin Releasing Factor or Hormone)[1] is a very rare substance produced in the hypothalamus, and seems to be of central importance in the mammalian endocrine system. The accepted view (whether constructed or discovered) is that TRF(H) triggers the relase of the hormone thyrotropin by the pituitary gland; this hormone in turn governs the thyroid gland, which controls growth, maturation, and metabolism.

The work on TRF(H) was done by rival groups led by Guillemin and by Schally, respectively. Each disputed the originality and significance of the other's work, but the Nobel Prize was awarded to them jointly in 1977 as co-discoverers. The sheer effort of the research work is something to be marvelled at. Guillemin had five hundred tons of pigs' brains shipped to him in Texas from slaughterhouses in Chicago. From this he managed to extract all of one mg of TRF(H). Schally has a similar story to tell about sheep's brains. Since there was so little to work with, standard tests on TRF(H) could not be done. It is the lack of any standard test that is at the heart of the problem; it is this which gives credibility to Latour and Woolgar's claims.

Contrast TRF(H) for a moment with gold. This precious metal, unlike the hormone, is clearly observable stuff.[2] Samples can be recognized as such by most people. However, mistakes and outright fraud are possible, so various assays were developed. These are tests which distinguish gold from other substances. In developing such assays standard samples of gold were available; they provided the means whereby one could test the test. If a particular assay could not distinguish gold from lead, then this failure would be readily apparent.

None of this can be said about TRF(H) for the simple reason that there was no recognizable sample. The plausibility of Latour and Woolgar's claim is based on this. Different bioassays were possible, and were considered by those working on TRF(H), but without a standard sample of the hormone there could be no test of the test itself. There is no way of telling which bioassay is the 'right' one, that is, gives us the right analysis of TRF(H). At least this is what Latour and Woolgar argue. So why was one particular bioassay chosen over any of its rivals? The answer comes from the sociology of science. The final choice of bioassay was the result of a kind of social or political negotiation, or, as they put it, the microprocessing of facts.

The 'fact' that all the fuss has been about is this: There is a substance in the hypothalamus that releases the hormone thyrotropin from the pituitary, and its chemical structure is pyroGlu-His-Pro-NH_2.

The second ingredient in their argument is this claim: 'Without a bioassay a substance could not be said to exist' (Latour and Woolgar 1979:64). There is no separate argument for it; the assumption just comes as a plausible part of the whole story. By putting the two parts together we easily get Latour and Woolgar's conclusion that facts are social constructs:

> TRF(H) exists if and only if bioassay B is accepted.
> B is accepted as the result of social negotiation.
> ∴ TRF(H) is not discovered, but is rather a social construction.

Neither premiss is acceptable. Consider the first, which implies 'Without a bioassay no substance can be said to exist.' Does this mean, for instance, that there was no gold until there was a test for it? Or that there was no Universe until there were thinking beings?

On the other hand, the expression 'can [not] be said to exist' might just mean that we have no right to *assert* the existence of any substance without a bioassay (or some other criterion) for it. This is much more plausible, as it leaves open the possibility that the substance does exist independently of us. But then, recalling the distinction made at the outset of this section, the bioassay only gives us the right to 'believe there is a fact'; we could be wrong. It follows that there still might or might not be a fact about TRF(H), and our bioassay might or might not be discovering it.

The second step in the argument suggests a picture of facts and procedures that we shall see again when discussing Collins below. The claim is that there is a tight little circle, that facts about TRF(H) and a particular bioassay stand or fall together. There is much that is right about this picture. What we believe about TRF(H) is highly dependent on a bioassay in a way that what we believe about gold is not. This is part and parcel of the network view of knowledge promoted as often as not by philosophers of science. Sociologists are extremely good at making plain the network character of our beliefs; the intricacies revealed are often staggering. Nevertheless, when it suits their purposes, the networks seem to become vanishingly small. Instead of TRF(H) being linked to all sorts of elements in our larger 'web of belief' (as Quine likes to call it), it is linked exclusively by Latour and Woolgar to a particular bioassay; and that bioassay in turn seems to be linked to nothing at all other than TRF(H).

Looking at Latour and Woolgar's own account, we can begin to see some of the possible connections in a bigger network. In the particular bioassay adopted, rats are used instead of mice because mice are thought to have more sensitive thyroids; males are used because it is thought the female reproductive cycle might interfere; the rats used are about 80 days old because it is thought that at that age the thyrotropin content of the pituitary is greatest; etc., etc. I do not wish to downplay the enormous number of factors taken into account; however, I do want to point out that each feature of the bioassay seems to be getting some sort of justification. It may be a fallible justification, but it is there none the less.

Latour and Woolgar are unimpressed with this sort of consideration. Not satisfied with the social construction of facts, they want to say more.

Our argument is not just that facts are socially constructed. We also wish to show that the process of construction involves the use of certain devices whereby all traces of production are extremely difficult to detect.

(Latour and Woolgar 1979:176; their italics)

Have we seen this before? Creationists sometimes claim that not only did God make the world a mere six thousand years ago, but, in order to test our faith, he also made the fossils to make things look very much older.

Let me turn my attention now to some of the themes of another anthropologist of the laboratory, Harry Collins.

TACIT KNOWLEDGE

Collins thinks there are two models of knowledge. In the 'algorithmical' version even a computer could be programmed to carry out various practical applications of propositional knowledge. He points out how silly such a view is (thereby demolishing the rationalist view, as he sees it) by simply noting that no one could read a book on mechanics and then get on a bicycle for the first time and ride it. But, of course, this is only a caricature of how rationalists view the activity and results of science; so some of us might be pardoned for remaining sceptical about Collins's treatment of these issues.

Collins's preference is for an 'enculturational model', illustrated by our ability to ride a bike. He makes much of the fact that this includes a large measure of 'tacit knowledge'. In the first of his case studies, Collins argues that the ability to replicate the TEA-laser[3] is a skill acquired, not from reading journal reports, but from on-site interaction with the builders of the original. He is always interesting and insightful, but not entirely convincing on this score, and he provides the reasons himself for why we should not believe him.

The original TEA-laser work was done by a Canadian military research establishment and was classified for two years. Later research groups working on the TEA-laser acted as rivals, not sharing information in a wholly generous way. Collins quotes one as saying: 'If someone comes here to look at the laser the normal approach is to answer their questions, but . . . although it's in our interests to answer their questions in an information exchange, we don't give our liberty.' Another of Collins's scientists remarked,

'Let's say I've always told the truth, nothing but the truth, but not the whole truth' (1985:55). In such an atmosphere, the notion of 'tacit knowledge' seems quite beside the point. These are secrets. It may well have been that a simple description would have allowed easy replication of the TEA-laser, if only one were given.

Why is some research classified? Why do research groups compete with one another and only reluctantly share information? Without a doubt the explanation must be in terms of social forces. Nevertheless, this has nothing to do with the content of the TEA-laser theory. There is a clear difference between social forces influencing a scientist to believe something and influencing that same scientist to keep secret something she already believes; it is a distinction Collins seems unwilling to make.

Perhaps, though, the fault is just with the example. (Collins is much more persuasive when describing one particular scientist's attempts to replicate his own laser than when describing one group's efforts to copy the achievements of another.) At any rate, I am perfectly prepared to agree with him that there are cases where propositional knowledge is not enough. Journal reports are not sufficient; we need hands-on experience under the guidance of masters. Skill picked up this way constitutes tacit knowledge and Collins rightly asserts its existence. However, he draws the wrong conclusions.

Only his sociological account, the 'enculturational' model of science, will allow for this sort of transmission of knowledge, says Collins (1985:57f.). But all of this is doubly misleading. First, there is nothing in the rationalist outlook which bars the existence of acquired skills. And second, the mere existence of tacit knowledge implies nothing social. Imagine a solitary Robinson Crusoe-like character who has never been a part of any human society. Could he not make a bike and learn to ride it on his own? If he could make and ride such a thing, then he would be replicating a common phenomenon (even though he would undoubtedly not conceptualize it as we do). Moreover, and this is the crucial issue, his knowledge would be just as tacit as ours, yet there would not be a shred of the social about it. If this example seems too far-fetched, then just consider rock-throwing skills. Our solitary character would have tacit knowledge of how to throw a rock, but there is nothing social about his skill.

The whole subject of tacit knowledge requires considerable

caution. To start with, 'tacit knowledge' is very much a success term. If I tacitly know how to ride a bike then I *can* ride a bike. If anyone has tacit knowledge of how to build a TEA-laser, then that person *can* build one. Let's suppose for a moment that the quantum theory of radiation is false. Then there simply aren't any Light-Amplification-by-Stimulated-Emission-of-Radiation devices. That is, there aren't any lasers – though, of course, there are things called 'lasers'. In such a case no one would know (tacitly or otherwise) how to build a TEA-laser after all. Ascribing tacit knowledge to someone presupposes a lot of background beliefs. We would not now, for instance, ascribe tacit knowledge to Priestley of how to work with phlogiston, nor to Blondlot with regard to N-rays. Tacit knowledge, even if it does have a significant social dimension as Collins suggests, is highly dependent on a theoretical background which has nothing particularly social about it.

The relativist-minded, such as Collins, may want to say that Preistley and Blondlot *did*, nevertheless, have tacit knowledge, tacit knowledge with respect to their frameworks. Fair enough, I don't want to argue against relativism here (although I certainly think it wrong). But note that the ascription *is* framework-dependent, there being no such thing as pure tacit knowledge. For rationalists and sociologists alike, then, tacit knowledge is a highly derivative notion. Even if it were intimately connected with social factors – which it isn't – it still couldn't carry the load sociologists like Collins want it to bear.

THE EXPERIMENTERS' REGRESS

There are two big themes in Collins's *Changing Order*. Tacit knowledge is explored in the study of the replication of the TEA-laser. The other is a problem Collins calls 'the experimenters' regress'. It arises when there is some question about the very existence of the phenomenon in question. If we have successfully built a laser, we are easily convinced that we have done so since we can zap things with it. Analogously, getting from A to B in one piece is a good sign that we know how to drive a car. Such relatively clear criteria of achievement are not always available, and that is how the epistemological problem arises.

Question: How do we know we have successfully built a
 gravity wave detector?

Answer: By the successful detection of gravity waves.
Question: But how do we know that these are gravity waves
    that have been detected (and not just noise)?
Answer: Because they have been detected by a properly
    working gravity detector.

Clearly this is question-begging. It is also remarkably akin to the view of Latour and Woolgar embodied in their claim that the existence of a substance is tied to the acceptance of a bioassay. Collins calls it 'the experimenters' regress' (Collins 1985:83f.), and he wonders how it is possible to break into the circle.

As Collins rightly notes, in such situations we can expect considerable disagreement among interested parties as to what has been achieved. Appraisals by scientists of each other's work range from admiration to contempt. None of this is really a surprise, but still it is nicely illustrated in *Changing Order*, especially by the case study on gravity waves. After clearly laying out a spectrum of opinion, Collins (rather hastily) concludes that 'there is, then, no set of "scientific" criteria which can establish the validity of findings in this field. The experimenters' regress leads scientists to reach for other criteria of quality' (1985:88).

Collins summarizes his view of experimentation in a table (1985: 146). Judgements of competence/incompetence of any particular experiment are said to depend on belief or disbelief in the phenomena being investigated.

		Scientist believes in phenomenon under investigation	
		Yes	*No*
Experiment finds results consonant with phenomena	Yes	Competent	Not competent
	No	Not competent	Competent

But this is rather simplistic as well as being downright wrong, and it is even at odds with Collins's examples. The scientists who rejected Weber's results still believed in the existence of gravitational waves – their existence is implied by an accepted theory, General Relativity. They just denied that Weber was able to measure them. And even though it was universally thought that Weber had failed to detect gravity waves, he was very far from being universally thought inept. The concept of 'experimental competence' is vastly more complicated than Collins allows.

In his detailed discussion of the case Collins is more subtle. When it comes to grand conclusions, he makes quite unwarranted leaps. The debate among rival scientists, as Collins sees it, is fought on two fronts represented by two questions: Is the experiment to detect gravity waves competently done? Do gravity waves exist? Collins is quite enlightening on the issue and rightly stresses the connection between the two; but he goes well beyond making this reasonable and important point when he further claims: 'Thus, the definition of what counts as a good gravity detector, and the resolution of the question of whether gravity waves exist, are congruent social processes. They are the social embodiment of the experimenters' regress' (1985:89).

In what significant sense are these processes social? It's a quite unwarranted jump from the fact that the debate is not easily resolved to the conclusion that the process is a social one. Perhaps the best way to defeat such a view is to do two things: One is to offer a plausible alternative account of how such debates are resolved; the other is to attack Collins's view directly. I'll very briefly try to do both.

The rationalist alternative to Collins's account involves stressing the theory-ladenness of observation and of judgement in general. We have a large number of background beliefs which will tell us what gravity waves are and how they can, in principle, be detected. Moreover, these various background beliefs will also tell us about the nature of the detecting device, how it works, its degree of sensitivity, the conditions under which it might fail, and so on.[4] This breaks the circle that Collins calls the experimenters' regress. It is not the detection of gravity waves (nor the belief that they do not exist) that uniquely determines whether the detector works; rather it is our well-grounded background beliefs which determine what has or has not been observed.

The application of these background beliefs is no simple matter. But we haven't abandoned all rationality just by abandoning what Collins calls the algorithmic model. Moreover, the background beliefs of rival scientists may even differ from one another significantly. When this is the case, the debate about the detection of gravity waves will be postponed and interest will focus to some degree on the rival background beliefs themselves. The process is extremely complicated and will often be long drawn out. But the crucial point is this: the circle can be broken by rational considerations; social factors are not needed to terminate the experimenters' regress.

Collins, like so many other sociologists of science, repeatedly endorses a *network* view of belief (drawn from Quine and Hesse): our beliefs and our concepts are not isolated, but are linked to every other belief and concept that we hold. Given this view, the pair 'Gravity waves exist' and 'This detector works' are not an isolated pair of beliefs bound in their own little circle, a little network consisting of just two sentences. They are undoubtedly close to one another in the whole network of belief (Collins is very good at making this clear), but the acceptance or rejection of each is not uniquely dependent on the acceptance or rejection of the other. The rest of the network can have a significant or even overwhelming influence.

Of course, sociologists are not going to be too sanguine about background theories, the other elements of the network; but that is a different matter. The point, I repeat, is only that such background theories do exist and can be employed to block the regress. Now Collins is not oblivious to these sorts of considerations. Apart from expressing qualms about the status of the background theories themselves, he would perhaps claim that there cannot be any uniquely successful or correct application of these background theories since the 'algorithmic model' is hopeless. Perhaps at this point we have reached an impasse, with the rationalist countering that complexity and difficulty are one thing, impossibility another.

So much for a rationalist alternative to Collins's account of the regress. What does Collins himself say about breaking into the circle? He posits a social entity, 'the core set', a group of people 'bound only by their close, if differing, interests in the controversy's outcome' (1985:142). Furthermore,

> The core set 'funnels in' social interests, turns them into 'non-scientific' negotiating tactics and uses them to manufacture certified knowledge. If one looks very closely, one can see how the outcome of core set debates is affected by these 'socially contingent' factors, but one can also see how the output is nevertheless what will henceforward be proper knowledge. *The core set gives methodological propriety to social contingency.*
>
> (Collins 1985:144)

Collins's model straightforwardly suggests that no one makes a contribution in a specific field unless that person is a member of the core set of that field. Counter-examples come to mind. To take but one, the great British mathematician G. H. Hardy had no interest at all in population genetics and could not be reasonably counted among any biology core set. Yet when he encountered a certain line of reasoning which was common among population geneticists, his good mathematical sense balked at it. He quickly worked out the now central Hardy–Weinberg Law. The law was quickly accepted by the core set after being forced upon them by someone with no interest in the specific outcome of their debates. To say the genetics community accepted the result because they had good reason to is obviously the right thing to say. To say the acceptance of Hardy's result was due to various social interests funnelled in by the core group, though logically possible, seems too implausible to be given any credence. It is not that there aren't any social factors at work, or that people don't have interests. Rather, it is that there are considerations which frequently arise, such as Hardy's little bit of mathematics, which completely overwhelm those social interests.

DOES THE WORLD MATTER?

Readers of Collins, and of Latour and Woolgar, will have the nagging suspicion that on their views the physical world counts for nothing in the production of scientific belief. Judgements of experimental competence are linked to judgements of the existence of phenomena; both are divorced from the world (the physical world, that is, not the social world). In support of this Collins, perhaps, has chosen his examples with care. Replication in each of his three case studies – lasers, gravity waves, and parapsychology – is, as he calls it, capricious. This fact may give the Collins thesis more support than it rightly deserves, since his persuasiveness may come from unusual cases.

At the time of my writing, the most exciting physics has to do with superconductivity at relatively high temperatures. Here replication (in many cases, though not all) has been a snap. Laboratories all over the world have been reproducing a wide variety of related phenomena within days of hearing reports on the original work of others. Even undergraduates (at least at the

University of Toronto) are making high-temperature superconductors as part of their standard laboratory education. Collins would have had a very much harder time trying to make his case with this example. The phenomenon of high-temperature superconductivity is both amazing and theoretically unexpected, and it will likely have enormous technological (hence social) ramifications. Nevertheless, *contra* Collins, the phenomena have been produced and reproduced with ease. Surely this is a case where scientists have been overwhelmed by the physical world, not by the social one.

This is a good place to inject a note of caution about the expression 'creating phenomena'. There is a potential for equivocation. Ian Hacking has recently been arguing that scientists often create new phenomena in the laboratory (see Hacking 1983). He means that scientists make things that never existed before. This sort of creation, however, is in no way social.[5]

Suppose scientists were somehow able to make a region of space free of any forces. They could then create, for the first time in the history of the universe, an inertial body. Obviously this inertial body is not a social construction. Its potential for existence has always been there, as embodied in Newton's first law: If there are no forces acting then the body will remain at rest or moving with uniform velocity in a straight line.

Similar things can be said about scientists who are at present creating all sorts of new high-temperature superconductivity phenomena. The ceramic materials used have never existed before, but the phenomena created are perfectly natural, nevertheless. Their potential for existence has always been, embodied in laws of nature which have a conditional form like Newton's first law. We just do not know, in this case, what those laws are.

A PARADOX

The fact that there is much to be learned from sociological studies of science presents something of a paradox for any rationalist: how can it be that an approach to science which is so wrong-headed nevertheless makes very valuable contributions to our understanding of science? Of course, we could deny that anything of value has been produced; then the problem does not arise. But it seems clear to me that there is much to be gained from reading either Collins or Latour and Woolgar; and this is true of numerous other sociological studies as well. Rationalists owe an explanation, so here is a brief attempt.

Behind the anti-rational rhetoric, sociologists of science are people of ordinary good sense. Couple this ordinary good sense with genuine observational skill and it is little wonder that something good results. Here is a case in point. At the beginning of Latour and Woolgar's *Laboratory Life*, very careful, 'unbiased' notes are kept by the 'anthropologist in the lab'. Among the things recorded are that the typewriters were noisy, that someone was eating an apple, and so on. But this pretence is kept up for only a page. Never again is someone's diet mentioned; nor is it stated why this sort of fact is no longer being recorded. Latour and Woolgar, just like Collins and just like most of the rest of us, have a pretty good feel for what is and what is not relevant in the doing of good science. The anthropologist of science would never get past the front door if this were not so, but would instead still be in the process of recording the infinitude of door-knob facts. We may not be able to say what the rules of good science are, but we generally recognize good science when we see it. Similarly, the anthropologist of the laboratory may not know the rules of English grammar – not even linguists do – but he can still write grammatical prose. So, in spite of themselves, sociologists of science (who are relatively skilled observers) often cotton onto factors which are, from a rationalist's point of view, quite relevant, and bring these to the reader's attention. That is why sociologists make contributions even a rationalist can value.

But, of course, this is intended to be little more than idle speculation.

A PARTING SHOT

In closing his book, Collins worries about the various social and political consequences of different attitudes to science. He takes a high moral tone toward the end of *Changing Order*:

> For the future *citizen* the [algorithmic] model of science and the natural world that developed through normal scientific teaching is positively dangerous for democracy and for the long-term future of science itself. The model allows the citizen only two responses to science: either awe at science's authority along with a total acceptance of scientists' *ex cathedra* statements, or

rejection – the incomprehending anti-science
reaction. . . . Where scientists' *ex cathedra* statements are found
wanting – as they inevitably will be from time to time – then the
most likely reaction is disillusion and distrust.

(Collins 1985:161)

Such an outcome, most of us would agree, is quite unpalatable.
While I'll leave the last word as to why it is unpalatable to Collins,
I will ask the reader to wonder what a thoroughgoing relativist
might mean when he says: 'A loss of confidence in the scientific
enterprise is a disaster that we cannot afford. For all its fallibility,
science is the best institution for generating knowledge about the
natural world that we have (1985:165).

Chapter Five

BOLINGBROKE VERSUS HENRY FORD

History is philosophy teaching by example. – *Bolingbroke*

History is bunk. – *Henry Ford*

If things appear to be so-and-so, then it is reasonable to take them to actually be this way, unless we have evidence to the contrary. Science appears to be a rational activity; indeed, it is our paradigm of rational activity. The preceding chapters have looked at alleged evidence to the contrary, but the allegations were found wanting. We are back, then, with common sense: the plausible thing to believe is that science is rational. This is not to say that it is entirely rational or even mostly rational, only that science is rational in some sense yet to be articulated.

The next question, quite naturally, is 'What is the nature of scientific rationality?' Before we can answer this question with any conviction we must know what would even count as evidence for or against any particular answer. It is at this point, I believe, that case studies have a role to play. Much has been made of case studies by the sociologists of science, but, I think, their approach has been systematically wrong-headed. However, my attention will be focused on a different target in this chapter since I find that there is almost as big a battle to fight with rationalist-minded philosophers on this issue.

Most rationalist philosophers have a schizophrenic attitude toward the history of science. On the one hand, they want their philosophical accounts to do justice to typical scientific practice; but on the other, fearing the so-called naturalistic fallacy, they want to avoid any 'confusion' of historical facts with philosophical norms. The standard rationalist view of historical case studies is much the same as the view of diagrams in a geometry book. Such pictures are heuristic guides which aid the imagination; they help us to understand the theorem and to follow the proof, but they play no evidential role at all. Their purpose is not logical, but psychological. Analogously, historical case studies are said to illustrate methodological norms; they inspire new ideas. But they do not justify; they are not evidence of any sort. This is the Henry Ford view of history: at worst it is bunk; at best, it is suggestive, but it is the sort of thing we can get on without.

Traditionally, the proposed norms of science have been argued for (or against) on conceptual, logical, and a priori grounds. Sociologists have sometimes had good reason for saying that rationalist-minded philosophers were simply out of touch with their own subject matter, real science. Case studies from the history (including contemporary history) of science have been brought forward by philosophers, if at all, as examples only, and used merely to illustrate a point; these episodes have never been brought forward as evidence. Usually philosophers of science have recommended the study of history only as a source of inspiration and discovery and never as a source of justification. The use of history has been insipid at best and often non-existent. The spirit of Henry Ford has been dominant.

Though I think the standard view of case studies among philosophers is wrong, it is, nevertheless, internally consistent. The problem with it arises when philosophers try to maintain it and at the same time concede something to Bolingbroke by insisting on 'doing justice to the history of science'. This schizophrenic attitude shines through clearly in the following two remarks by Wesley Salmon:

> If a philosopher expounds a theory of the logical structure of science according to which almost all of modern physical science is methodologically unsound, it would be far more reasonable to conclude that the philosophical reasoning had gone astray than to suppose that modern science is logically misconceived.
>
> (Salmon 1970:73)

But after conceding so much, he takes it back when he adds:

> In spite of this the philosopher of science is properly concerned with issues of logical correctness which cannot finally be

97

answered by appeal to the history of science. . . . [S]olutions, if
they are possible at all, must be logical, not historical in
character. The reason, ultimately, is that justification is a
normative concept, while history provides only the facts.

(Salmon 1970:74)

In the first passage, Salmon grants the evidential role of factual
history to normative methodology, but in the second he withdraws
it. It would seem that historical facts can be used negatively to
refute a philosophical theory, but they cannot be used positively to
confirm one. Only a priori arguments could establish, say, Popper's
falsificationism or Bayesian inductivism. However, this sort of
asymmetry in the historical evidence is not typical of Salmon's
thinking. He does not, for instance, think that experiments can
only refute and not confirm ordinary scientific theories, nor do
most other philosophers who share his attitude toward the history
of science.

The problem we are faced with is this: how are we to overcome
this ambivalence toward history? How could one avoid a
schizophrenic state of mind? There are, I think, two ways. One
way is to join Henry Ford and do methodology in a completely
a priori way. The most recent champion of the Henry Ford outlook
is Daniel Garber:[1] 'the history of *actual* scientific change is,
surprisingly enough, largely irrelevant to the *normative* theory of
scientific change' (Garber 1986:92). This is to claim, in consequence, that it is entirely possible that all actual scientific practice,
past and present, is irrational and 'unscientific', which is, in turn,
to accept the (I think, absurd) further consequence that scientists
might be bad at doing science.

The other way to cope with the problem is to claim that the
history of science is, in some significant sense, rational and that
history can be used as evidence for (or against) doctrines of how
science ought to be done. In short, it is to claim that scientists are
good at doing science and consequently that there is an evidential
relationship between the history of science and the normative
philosophy or methodology of science. This way is Bolingbroke's,
and I shall adopt it.

In taking this course I join a minority of historically oriented
philosophers of science who have a similar attitude toward the role
of case studies. Consider, for a moment, some important examples.
Thomas Kuhn has often been accused (perhaps with some justice)

of making the history of science come out irrational. He has long
protested against this claim, although unlike a priori methodologists, he admits the gravity of the charge.

> I do not for a moment believe that science is an intrinsically
> irrational enterprise . . . Scientific behavior, taken as a whole, is
> the best example we have of rationality. . . . [I]f history or any
> other empirical discipline leads us to believe that the
> development of science depends essentially on behavior that we
> previously thought to be irrational, then we should conclude not
> that science is irrational but that our notion of rationality needs
> adjustment here and there.
>
> (Kuhn 1971:144)

Imré Lakatos, who was so fond of paraphrasing Kant –
'Philosophy of science without history of science is empty; history
of science without philosophy of science is blind' (Lakatos
1971:91) – also overcomes the ambivalent attitude toward actual
scientific practice, by embracing its history as a source of evidence
for methodological norms. Refusing to believe science irrational, he
writes:

> Is it not then *hubris* to try to impose some *a priori* philosophy of
> science on the most advanced sciences? Is it not *hubris* to demand
> that if, say, Newtonian or Einsteinean science turns out to have
> violated Bacon's, Carnap's or Popper's *a priori* rules of the game,
> the business of science should be started anew?
>   I think it is . . .
>
> (Lakatos 1971:121)

Larry Laudan once held that there has been recently 'and
catastrophically, a growing tendency (especially in the English-
speaking world) to imagine that one could grapple with the nature
of knowledge while remaining blissfully ignorant of its best extant
example – the natural sciences' (Laudan 1977:1).

> [T]oo many discussions of scientific rationality and progress
> have been both uninformed by, and inapplicable to, the actual
> course of the evolution of science. Accordingly, historical cases
> and episodes will be used extensively in this essay [i.e., *Progress
> and its Problems*], these are intended not merely to *illustrate* my
> philosophical claims, but also to *test* them. If the model under
> discussion here fails to illustrate the manner in which scientific

decision making has actually worked (at least some of the time),
then it will have failed entirely in its ambitions.

(Laudan 1977:7)

It might be easy to find others of a similar sentiment; but
quoting *ad nauseam* would not make the point any more effectively.
The point is this: the past matters; history, *contra* Henry Ford, is
not bunk, and rather, as Bolingbroke insisted, it has something to
teach us. The history of science has some evidential role to play in
the evaluation of competing methodologies of science. Accepting
this is the best way of overcoming historical schizophrenia, a
malaise which afflicts far too many rationalists. It will be the point
of departure for the rest of this chapter, and the next.

Historically oriented philosophers have produced a great
abundance of case studies in recent years, and the message comes
through clearly that history is indeed being taken seriously. The
claim, though often only implicit, is that the historical case study is
more than an illustration, it has a real evidential role to play as
well. The trouble is that there has been very little discussion as to
how it is that historical examples do their evidential job. We need
to know how case studies work. It is a very significant step to join
Bolingbroke in saying that history teaches us philosophical norms,
but questions remain: how does history instruct? and what is being
taught?

One suggestion about the history–methodology evidential relationship which we should reject right at the outset is a
straightforward naturalism, as championed recently by Ron Giere.
He remarks that 'the *nature* of the relationship . . . is clear. It is the
standard relationship between theoretical models and empirical
data' (Giere 1984, p.28). I have no quarrel with aspects of this
proposal. Indeed, I think there is very likely a theory–data
interaction. But ultimately it won't do, and the reason is perfectly
simple: nature doesn't cheat, or make mistakes, or react to social
forces; scientists do. A theory of, say, gravity cannot allow the
existence of a single (real) counter-example, while the correct
theory of rationality can tolerate lots of irrational acts, just as the
existence of murderers does not refute our moral belief that murder
is wrong.

Of all the philosophers who follow Bolingbroke, that is, who
'take history seriously' and who try to 'do justice to actual science'
remarkably few have tried to say just how it is that history is to

serve as a way of sorting the philosophical wheat from the chaff. The views of Imre Lakatos, Larry Laudan, and a small handful of others pretty well exhaust the field. As we shall see below, their views have some definite shortcomings. It is of the utmost importance, therefore, that the situation be cleared up. If sense cannot be made of the Bolingbroke notion of history teaching us by example, that is, of somehow being evidence for methodology, then much of the foundation of historically oriented philosophy of science crumbles, because history then need not be taken seriously at all. Anti-historical philosophers, with some justice, will be able to steadfastly maintain that historical case studies *are* like diagrams in a geometry book. History, they can justly claim, is a psychological aid only, and plays no logical or evidential role whatsoever. Henry Ford will have won by default.

It is not enough merely to side with Bolingbroke. What is needed is an account of the history–methodology evidential relationship which is both clear and coherent and which specifies in detail the role of historical case studies.[2]

LAUDAN'S PRE-ANALYTIC INTUITIONS

In Larry Laudan's *Progress and its Problems* (1977), history is used in many significant but distinct ways. One use concerns the evaluation of a theory or a research tradition. That a theory belongs to a tradition with a long, illustrious past would and should count in its favour. A second use, the use which will be critically scrutinized here, is in the testing of methodologies: Laudan maintains that we have trustworthy intuitions about a number of episodes in the history of science and that these can be used to test competing methodologies:

> [T]here is, I shall claim, a subclass of cases of theory-acceptance and theory-rejection about which most scientifically educated persons have strong (and similar) normative intuitions. This class would probably include within it many (perhaps even all) of the following: (1) it was rational to accept Newtonian mechanics and reject Aristotelian mechanics by, say, 1800; (2) it was rational for physicians to reject homeopathy and to accept the tradition of pharmacological medicine by, say, 1900; (3) it was rational by 1890 to reject the view that heat was a fluid; (4) it was irrational after 1920 to believe that the chemical atom had no parts . . .

101

What I shall maintain is that there is a widely held set of normative judgments similar to the ones above. This set constitutes what I shall call *our preferred pre-analytic intuitions about scientific rationality* (or 'PI' for short) . . . [O]ur intuitions about such cases can function as decisive *touchstones* for appraising and evaluating different normative models of rationality . . . *[T]he test of any putative model of rational choice is whether it can explicate the rationality assumed to be inherent in these developments . . . [and] the degree of adequacy of any theory of scientific appraisal is proportional to how many of the PIs it can do justice to.*

(Laudan 1977:160f.)

What Laudan offers us is this: a set of judgements about episodes in the history of science that we clearly and distinctly comprehend. That is, we *intuitively* grasp that each of the events was rational (or irrational, as the case may be) and we know the rationality (or irrationality) of these episodes *before* (logically and perhaps temporally) we have written about them or reflected upon them from any particular normative point of view. (Of course a distinction must be made between the history of science, that is, the events, and the historiography of science, the written story. The PIs are a set of events.) Laudan's criterion of acceptance for a theory of scientific rationality is simply based on the degree to which it captures the PIs. And this just means that the historiography of the episodes in question, based on a particular theory of scientific rationality, has to make the episodes rational (or irrational) as judged by our pre-analytic intuitions.

It is only the PIs which are used for testing purposes; only the PIs are to be *described*. Laudan says ' . . . the philosophy of science is both descriptive and normative, both empirical and a priori, but with respect to different types of historical cases' (Laudan 1977:163). The rival methodologies are required to describe the PIs and prescribe the rest. Testing is done against the PIs, and the rest of the historical episodes outside the set of PIs are not used for testing purposes at all. This distinction is to avoid the naturalistic fallacy, but it leads to difficulties. Below I shall consider some variations of Laudan's account, modifications which he might well welcome, to see if they fare any better.

Laudan's account of the evidential relationship between history and methodology has a number of serious difficulties – enough, I

102

think, to render it implausible. For one thing, his proposal is fundamentally at odds with his model of scientific rationality. There are a number of reasons for this which I shall get to shortly, but in the main it is because the PIs act as a foundation in an anti-foundationalist theory of method. Furthermore, his account is ill-motivated since it largely stems from his efforts to skirt a problem of 'vicious circularity' which is, as I shall argue below, a mere pseudo-problem.

The actual account of rationality that Laudan proposes is to apply not only to science but to all activities that have cognitive aims. Thus, his model is to apply to itself, as I am sure he will readily agree. What I will argue is that his proposal for testing methodologies is in flagrant violation of that model itself. In order to show this I will first give a brief description of that methodology (as formulated in Laudan 1977).

According to Laudan it is not individual theories which are tested directly, rather, it is the *research tradition* which is the unit of appraisal. The aim of a research tradition is to solve problems, of which there are two kinds. *Empirical problems* arise for a theory when a competitor theory has explained some phenomenon. This point must be stressed. There is no such thing as an observation which is not theory-laden, nor is there a neutral observation language. A particular phenomenon falls into the domain of a theory when that theory or a competitor can explain it. A theory is required to explain it only when a competitor theory has managed to. Until then it is free to ignore the phenomenon in question, as it is unclear that the phenomenon is in any way relevant.

The other kind of problem, the *conceptual problem*, arises for a theory when it clashes with other accepted theories (or research traditions), or when it is internally inconsistent. Having unsolved problems, whether empirical or conceptual, counts against a theory or tradition, while solving them counts in its favour. Specifically, scientific rationality, according to Laudan, consists in following these rules: first *accept* the tradition with the *greatest momentary adequacy*, that is, the one which has solved the most problems and has the fewest anomalies hanging over its head; second, *pursue* the tradition which is the most *progressive*, that is, the one which is solving problems at the fastest rate.

When applying Laudan's model of rationality itself to the theory of how to test any model of rationality we need to know what the analogues of the conceptual and empirical problems are. It seems fairly easy to understand what a conceptual problem for a methodology might be. Internal inconsistency, of course, is one sort, and some commentators have expressed concern over Laudan's disavowal of 'truth' while others have doubted Laudan's ability ever to be able to individuate his 'solved problems'. Whether these are legitimate worries or not is of no consequence here; they illustrate the kinds of thing which could be called 'conceptual problems' for a normative philosophy of science.

But what is the analogue at this meta-level of the empirical problem? The only candidate would seem to be capturing the PIs. A methodology solves empirical problems by making the individual episodes in the set of PIs turn out rational or irrational as our intuitions demand. So, for example, if we intuitively think it was rational to accept Newtonian mechanics by 1800, then we require of any normative methodology that it say as much. Failure to do so would result in an evaluation similar to our appraisal of a biological theory which denied that grass is green. At any rate, I shall work on the assumption that this is the right way to understand Laudan's meta-methodology. Now to some difficulties.

First, according to Laudan's methodology, an empirical problem arises only when a competitor theory has explained a phenomenon. (In the case of Brownian motion, for example, phenomenological thermodynamics was under no obligation to explain it until the kinetic theory had done so.) It follows that theory evaluation is basically *comparative*. Yet Laudan's PIs function as an absolute measure; they are put forward as a *sine qua non* of adequacy before any other theory has managed to explain them. Even if no account of scientific method could capture them, the PIs would still serve as an absolute standard. Thus, Laudan's methodology (quite rightly, I think) calls for *evaluation by comparison* while his account of testing methodologies is, contrary to this, quite anti-comparative.

A second difficulty concerns the fact that the PIs function as a sort of *foundation for methodological knowledge*. Yet Laudan's model of rationality is highly anti-foundationalist. Nothing serves as a base, corrigible or not; he maintains a coherence view of justification. A consistent application of his model of rationality to the account of testing competing models would have to rule out any notion that the PIs were immune to revision. They might serve, like Popper's basic statements, as a convenient starting-point, but they might all be tossed out as time goes by. By stressing conceptual matters and the theory-ladenness of observation over 'the given' in his model of rational scientific change, Laudan rightly downplays those spurious entities, *empirical facts*. By postulating the PIs, however, he has made himself a slave when evaluating methodologies to equally spurious entities, *normative historical facts*. The existence of an incorrigible set of PIs constitutes a type of foundationalism which is in conflict with his totally coherentist methodology.

Let us look more closely at the notion of an intuition. The term 'intuition' has two distinct senses. Which does Laudan mean when he says that there is 'a subclass of cases of theory-acceptance and theory-rejection about which most scientifically educated persons have strong (and similar) normative intuitions' (Laudan 1977:160)? There is one sense in which to have an intuition is to have immediate knowledge of a concept where having this knowledge does not entail being able to explicitly define the concept. So, to say, that some particular historical episode is intuitively rational is to say that we know that it *is* rational but we may still not be able to say *why* it is rational or even what rationality itself is. If this is what Laudan means by 'intuition' then the PIs must, once again, be viewed as foundational. They are pure, unadulterated, non-theory-laden, normative historical facts. But, as already noted above, this is completely at odds with the spirit of his methodology.

The other sense of 'intuition' has to do with common sense, or, more to the point, common prejudice. Many of our beliefs have this character. They seem obvious only because they have been ingrained since birth. It might well be the case that a particular example, such as the rationality of adopting Newtonian mechanics by 1800, strikes us as intuitive simply because we have been told over and over again since early childhood that failing to adopt it constitutes the height of superstition and folly. If the PIs are intuitive in this sense then it might well be wondered why we should take them seriously at all. We might just as well pick them from a hat.

This may seem too harsh a judgement, because some philosophers have claimed a priority for common sense and maintained that we should trust common-sense beliefs until they are shown to be false. I'm sure this is the right thing to do. However, showing a common-sense belief to be false usually amounts to overruling it by a theory for which there is good independent evidence. Ordinarily this is fine, but here the common-sense beliefs are the only evidence any methodological theory must account for; and so no methodological theory could overrule that evidence. Whichever episodes we pick for the PIs, the rationality of these episodes will thus become permanently entrenched common sense. Accordingly, Laudan's 'intuitions' are certainly suspect, and they provide a poor criterion for selecting examples of good and bad scientific practice to test competing methodologies.

Third, Laudan, quite rightly, stresses the great importance of history to the philosophy of science. The general aim of his whole programme is one of making explicit the many complex and widespread interrelationships between history and philosophy. In view of this it is surprising to find that most of the episodes in the history of science will be evidentially neutral to normative philosophy of science. The reason for this is very simple. The set of historical episodes is large while the set of PIs is rather small. Since 'the degree of adequacy of any theory of scientific appraisal is proportional to how many of the PIs it can do justice to' (Laudan 1977:161), (that is, since it is only the PIs which matter) most of history is evidentially irrelevant.

This is so contrary to the spirit of Laudan's general programme that there is a strong temptation to think that, if queried on the matter, he would put things somewhat differently and add additional criteria to the requirement of capturing the PIs. One possible response to this is to try to enlarge a set of PIs. Perhaps, as time passes, we could add more episodes. But on what basis? A new criterion for indefinitely enlarging the set of PIs would be a criterion for distinguishing good science from bad; that is, it would be a methodology. But, finding such a criterion is, in fact, equivalent to solving our initial problem.

Fourth, one of the great insights of Laudan, and of all of the members of the group of historically sensitive philosophers of science, is that scientific theories are evaluated by several (rational) considerations other than mere empirical data. Laudan himself stresses the great importance of conceptual problems; the fit, or lack of it, with other theories is often of paramount importance in theory appraisal. Accordingly, it is somewhat surprising to see him make history, and only history, the test for rival methodological theories. If a priori, conceptual, and logical considerations are rational ingredients in, say, physics, why not have such a priori,

conceptual and logical considerations playing a role in methodological appraisal? It is one thing to make historical considerations
important; it is another to make them exclusive. There is a certain
irony here. Many a priori methodologists, such as the positivists,
made themselves slaves to empirical facts; Laudan, after decrying
such a limited method of evaluating scientific theories, makes
himself a slave to historical facts in the appraisal of normative
theories.

Laudan has recently (1984, 1986) changed his views on this issue
considerably, dropping the PI criterion in favour of a version of
naturalism. I will criticize the new view below. It has been worth
going through his older view in detail, since it has been a major
contribution to the issue and will doubtless remain influential. I
will also postpone until later my remarks on an alleged problem of
circularity which seems to have motivated Laudan to propose the
PI criterion in the first place. It will be much more convenient to
take up the circularity issue below. I should note in passing, and I
stress this, that my criticisms of Laudan's criterion are partly based
on his own model of rational scientific change. It is a methodology
which I find appealing, and I think is left more or less untouched
by my criticisms of his meta-methodology.

LAKATOS'S RATIONAL RECONSTRUCTIONS

Imré Lakatos, as I mentioned above, was fond of paraphrasing
Kant: 'History of science without philosophy of science is blind:
philosophy of science without history of science is empty' (Lakatos
1971:91). He certainly practised what he preached in the sense that
all of his philosophical work was highly historical in nature, and
his historical case studies (whether scientific or mathematical) are
quite unlike anything a non-philosophical historian would even
dream of producing. As ideologue and example Lakatos was, until
his premature death in 1974, one of the main driving forces in
historical philosophy of science.

Lakatos accepted the fact that history had an evidential role to
play in choosing a methodology. Indeed, he complained that any
sort of methodology which is ahistorical and arrived at in an
a priori fashion will fail to do justice to the history of science.[3] Let
us look now to his specific proposals. Lakatos says in an essay on
the evidential relationship that he will

show that methodologies may be criticised without any direct reference to any epistemological (or even logical) theory, and without using directly any logico-epistemological criticism. The basic idea of this criticism is that *all methodologies function as historiographical (or meta-historical) theories (or research programmes) and can be criticised by criticising the rational historical reconstructions to which they lead.*

(Lakatos 1971:109)

His account of the normative methodology testing procedure is similar to his account of the testing of scientific theories.

This normative-historiographical version of the methodology of scientific research programmes supplies a general theory of how to compare rival logics of discovery in which (in a sense carefully to be specified) *history may be seen as a 'test' of its rational reconstructions.*

(Lakatos 1971:109)

Lakatos's suggestion is that the best normative methodology is the one which will rationally reconstruct the history of science in such a way that it captures 'the basic appraisals of the scientific elite' (1971:111), that is, the value-judgements of the best scientists of the past.[4] In order to critically discuss his account it will be best to focus on certain key notions such as 'rational reconstruction' and 'basic value-judgement'.

Rational Reconstructions

Lakatos, in a monograph-length essay, makes the following methodological claim for the study of history:

In writing a historical case study, one should, I think, adopt the following procedure: (1) one gives a rational reconstruction; (2) one tries to compare this rational reconstruction with actual history and to criticize both one's rational reconstruction for lack of historicity and the actual history for lack of rationality. Thus any historical study must be preceded by a heuristic study: history of science without philosophy of science is blind.

(Lakatos 1970:138; his italics)

He then proceeds to give rational reconstructions of two research programmes: Prout's concerning the atomic weights of the elements and Bohr's concerning light emission and the quantized atom.

The term 'rational reconstruction' as it is usually employed in philosophy means, roughly, a re-description. Something which is described or accounted for in a loose and informal fashion is rationally reconstructed when it is described or accounted for using the precise and technical terms of some theory. Thus, an argument in a natural language may be rationally reconstructed by translating it into the symbols of formal logic. One of the things we would want to preserve in such a rational reconstruction is the validity or the invalidity of the initial argument. (Indeed, we often translate or reconstruct to test the argument for this very property.) Analogously, it would seem desirable to preserve the historical facts and the cogency of the reasoning in a rational reconstruction of an episode in the history of science. To the eternal consternation of most readers, Lakatos's rational reconstructions do not seem to do this at all.

> Thus in constructing internal history the historian will be highly selective: he will omit everything that is irrational in the light of his rationality theory. But this normative selection still does not add up to a fully fledged rational reconstruction. For instance, Prout never articulated the 'Proutian programme': the Proutian programme is not Prout's programme. *It is not only the ('internal') success or the ('internal') defeat of a programme which can only be judged with hindsight: It is frequently also its content.* Internal history is not just a *selection* of methodologically interpreted facts: it may be, on occasions, their *radically improved versions*.
>
> (Lakatos 1970:106)

As for the differences between the old history and the new, improved version: 'One way to indicate discrepancies between history and its rational reconstruction is to relate the internal history *in the text*, and indicate *in the footnotes* how actual history 'misbehaved' in the light of its rational reconstruction' (1970:107). The reason this rational reconstructing has to be done, says Lakatos, is because '*History without some theoretical "bias" is impossible*' (1970:107). But it is not in the least clear what this means. Fred Suppe remarks that Lakatos maintains 'the questionable thesis that all history of science *must* reconstruct history from the perspective of *some* scientific methodology' (Suppe 1977:669). He continues,

This appears to imply that unreconstructed history is
impossible – which seems to be incompatible with Lakatos's own
practice of indicating how his own reconstructed histories
deviate from the actual (unreconstructed) history by presenting
the *actual* history in footnotes. Making coherent sense of his views
here is difficult, if not impossible.

(Suppe 1977:669n.)

At the end of the essay in which he gives his account of the
evidential relationship between history and philosophy Lakatos
flippantly remarks:

Let me finally remind the reader of my favourite – and by now
well-worn – joke that history of science is frequently a caricature
of its rational reconstructions; that rational reconstructions are
frequently caricatures of actual history; and that some histories
of science are caricatures both of actual history and of its
rational reconstructions. This paper, I think, enables me to add:
Quod erat demonstrandum.

(Lakatos 1971:122)

It does not seem likely that Lakatos's rational reconstructions
could play a role in the appraisal of competing normative methodologies of science. The general reaction to his notion of a rational
reconstruction has been not merely one of scepticism but of
downright incredulity (even from philosophers who are historically
sensitive). For instance, Thomas Kuhn complains:

What I am trying to suggest, in short, is that what Lakatos
conceives as history is not history at all but philosophy
fabricating examples. Done in that way, history could not in
principle have the slightest effect on the prior philosophical
position which exclusively shaped it. . . . When one's historical
narrative demands footnotes which point out its fabrications,
then the time has come to reconsider one's philosophical
position.

(Kuhn 1971:143)

Gerald Holton is not too pleased with Lakatos's treatment
either; he considers it 'an ahistorical parody that makes one's hair
stand on end' (Holton 1974:75). And even though Laudan's book
Progress and its Problems is much influenced by Lakatos, still, he has

no sympathy for the latter's rational reconstructions, which he takes to be 'consciously and deliberately falsifying the historical record' (1967:170).

All the critics of Lakatos I have cited are historically orientated; they do not have anti-historical blinkers on. If anything they would be predisposed to be sympathetic. So, if Lakatos is not guilty of proffering false doctrine, he is at least guilty of presenting it in an extremely confusing way. While the authority of numbers does not count as evidence for falsity, it does count when the issue is clarity. If everyone finds a doctrine confusing then it is confusing! Nevertheless, I am inclined to think there is something deeply insightful in Lakatos's account of the history–methodology relationship. Something akin to his rational reconstructions plays an important role in the account of the evidential relationship which I give below. This is my second reason for quoting so extensively from Lakatos's critics. Citing those complaints here helps to set the stage for my own views presented in the next chapter.

Basic value-judgements

What Lakatos seems to want to capture with his rational reconstructions are those episodes in the history of science which are acknowledged to be examples of great science. For instance, he says:

> While there has been little agreement concerning a *universal* criterion of the scientific character of theories, there has been considerable agreement over the last two centuries concerning *single* achievements. While there has been no *general* agreement concerning a theory of scientific rationality, there has been considerable agreement concerning whether a particular single step in the game was scientific or crankish, or whether a particular gambit was played correctly or not. A general definition of science thus must reconstruct the acknowledgedly best gambits as 'scientific': if it fails to do so, it has to be rejected.
>
> (Lakatos 1971:111)

This amounts to the claim that basic value-judgements can be used to test competing rational reconstructions in just the same way as basic statements can be used to test competing scientific theories.

111

At first sight this seems to be the same sort of criterion for testing methodologies as Laudan proposed.[5] In so far as it is, it is open to the same objections.

The two criteria are not exactly alike, however, for they differ in one important respect. Lakatos's intuitions or basic value-judgements (which rational reconstructions are trying to capture) are not just *ours* (i.e., philosopher-historians here and now), as they are with Laudan, but, rather, they are the intuitions of the historical participants, the practising scientists, themselves. It is the basic value-judgements of Newton, Galileo, and Darwin, not us, which matter. Accordingly, we cannot charge Lakatos, as we did Laudan, with using suspect data, which might merely reflect *our* present-day prejudices. (If prejudices, then they are of long standing.)

Does Lakatos's account fare any better than Laudan's? Feyerabend (1975) does not think so, for two reasons. One is that by taking basic value-judgements of scientists, Lakatos has simply begged the question against his (Feyerabend's) general critique of science. He asks rhetorically: Why not take the basic value-judgements of astrologers or witch doctors? His second objection to Lakatos concerns the practical utility of the approach. In particular, Feyerabend claims, the basic value-judgements of scientists are simply not as uniform as Lakatos thinks they are and requires them to be for the purposes of testing norms. Feyerabend notes:

> Science is split into numerous disciplines, each of which may adopt a different attitude towards a given theory and single disciplines are further split into schools. The basic value judgements of an experimentalist will differ from those of a theoretician (just read Rutherford, or Michelson or Ehrenhaft on Einstein). . . . Even individual scientists arrive at different judgements about a proposed theory: Lorentz, Poincaré, Ehrenfest thought that Kaufmann's experiments had refuted the special theory of relativity and were prepared to abandon the relativity principle in the form proposed by Einstein while Einstein himself was of a different opinion.
>
> (Feyerabend 1975:202)

How Much is Rational?

Lakatos connects his methodology of scientific research programmes with the notion of rational reconstructions. This is done in stages. First he entertains a Popperian-type criterion: 'let us propose tentatively that if a [theory of rationality] is inconsistent with the basic appraisals of the scientific elite, it should be rejected' (1970: 111). Then he notes that this is too strong. It would rule out every methodology, and so it must be suitably modified. The way the meta-criterion should be changed parallels Lakatos's own modification of Popper's falsificationism for scientific theories.

> I emphasize the far-reaching analogy between scientific and methodological research programmes and stress that just as empirical 'basic statements' can be overruled by theory normative 'basic judgements' can be overruled by methodology.
>
> (Lakatos 1971:180)

Just as progressiveness is the key to appraising scientific research programmes, so also progressiveness is the key to appraising methodological research programmes. In neither case is there an incorrigible foundation. The sequence which contains first Popper's and then Lakatos's scientific methodologies is a series which Lakatos sees as progressing. And just what is the sign of progress? '[A] good rationality theory', says Lakatos, 'must anticipate further basic value judgements unexpected in the light of its predecessors or that must even lead to a revision of previously held basic value judgements' (1971:117). And furthermore, 'progress in the theory of scientific rationality is marked by discoveries of novel historical facts, by the *reconstruction of a growing bulk of value-impregnated history as rational*' (Lakatos 1971:118: my italics).

For Lakatos, progress seems to come to this: a methodological re-search programme is progressing if it is turning ever more episodes in the history of science into examples of rational scientific activity and, consequently, overruling fewer basic value judgements.

There is one grave problem with Lakatos's account on this interpretation, though. It will eventually make too much rational. It is certainly a virtue of a scientific theory to capture ever more empirical observations. But it need not be a virtue for a theory of rationality to make ever more episodes in the history of science turn out rational. The analogy between scientific research programmes and methodological research programmes breaks down here. Nature is infallible; it does not lie, cheat, or make mistakes. Human scientists do. The intuitions of the scientific elite cannot be always right. Progress for a scientific research programme is (among other things) toward complete empirical adequacy. The analogue of this for methodological research programmes, complete rationality of *all* episodes in the history of science, is absurd. Sociologists are not alone in thinking that silly. We know all too well, in advance of any serious investigation, that some episodes are irrational. We need not know which ones are irrational, but we most certainly do know that examples of bad science exist somewhere. The point was made by R. J. Hall, who rightly protested:

> Nobody would want to say that *all* of science is rational and that *all* of scientists' judgements about science are correct. We all recognize that there are Lysenkoesk episodes in the history of science – episodes that nobody would require a methodology of science to show to be internal and rational (scientifically rational, that is).

> (Hall 1971:157)

In sum, a number of objections have been raised against Lakatos here. His notion of a rational reconstruction is certainly a confusing doctrine, if not downright false. His reliance on intuitions or 'basic value judgements' is suspect, and subject to the same sorts of criticism as Laudan's account and to some more besides. And finally his notion of progressiveness in methodological research programmes leads to the absurd prospect of the whole of the history of science coming out rational. Lakatos's account of the history–methodology evidential relationship is simply unacceptable. As well as these criticisms, we should further note that Lakatos simply put his criterion for testing rival methodologies forward with hardly a word of explanation, and none of justification. So, one might well wonder why should we even consider accepting it anyway?

However, there are many good ideas in this confused mess, and in the next chapter I shall put forward an account which follows Lakatos in certain respects. In what ways our two accounts coincide is hard to say, since Lakatos is so unclear about what his rational reconstructions are. I do not wish to become embroiled here in exegetical quarrels, so I will simply say only that my account is inspired by his. I should also stress again, as I did after discussing Laudan, that it is not Lakatos's model of rationality that is here under attack. It is his proposal for testing models of rationality. Both Laudan and Lakatos claim that the method of appraising scientific theories should be the same as the method for testing methodological theories. This may seem natural at first blush, but there is really no compelling reason for this assumption. I find both of their models of rationality appealing and I further think that these methodologies are largely untouched by the foregoing criticisms.

REFLECTIVE EQUILIBRIUM

In *A Theory of Justice* (1971), John Rawls proposed a method for evaluating ethical theories known as *reflective equilibrium*. Though Rawls was only interested in ethics, the method is obviously a plausible way of evaluating *any* kind of normative theory. I shall briefly examine the method, suitably adapted for the history–methodology relationship.

Rawls's account of the process of justifying an ethical theory is quite simple and straightforward.[6] We start with individual considered judgements (for example, slavery is wrong, truth-telling is right), then we construct general principles (for example, utilitarianism, contractarianism) which are to account for these. Sometimes we will hold fast to the considered judgement and revise the general principles when there is conflict. At other times we will find the principle sufficiently satisfactory that a conflicting judgement will be revised rather than allow the general principle to be abandoned. We should do whichever is appropriate. This process continues until we reach reflective equilibrium, which is a point of (temporary) stability. (There is no permanent stability; new facts and new considered judgements will arise, which lead to further modifications.)

It is a simple matter to modify Rawls's method for ethics into a method for justifying normative principles for science. The result will be similar to, but will differ in important respects from, Laudan's method discussed above. First we should start with considered judgements about individual cases in the history of science (for example, Newton was rational, Lysenko was not). Then we would propose general methodological rules (for example, Bayesian

Inductivism, Popperian Falsificationism) which try to account for these considered judgements. When there is conflict between the considered judgements and the general principles, we will sometimes modify the one, and sometimes the other; we should do whichever is appropriate in the circumstances. We continue this process until we have reached a state of equilibrium. William Newton-Smith briefly describes a process similar to this (1981:209), and even uses an analogy similar to one drawn from linguistics that Rawls employs.[7] Perhaps he can be seen as a real exponent of reflective equilibrium in meta-methodology.

The important difference between Laudan and Rawls centres on the nature of the intuitions. Rawls's are revisable, while Laudan's are not. Laudan's pre-analytic intuitions can be overruled, but they remain intuitive nevertheless. The result of overruling one is forever to be stuck with something which is counter-intuitive. The considered judgements of Rawls, on the other hand, are completely corrigible, so that even the opposite of any one of them can later become a considered judgement. (Incidently, the intuitions of Lakatos's scientific elite are not revisable either. Their incorrigibility is not because they are *intuitions* of the past, but because they are *past* intuitions.)

The non-foundationalist nature of these intuitions has led Laudan (in a private communication) to explicitly reject reflective equilibrium as a suitable meta-methodology. He thinks it will lead to an inevitable self-supporting circularity. I consider this sort of worry about circularity later (see p.145), and show that it is not really a problem. Nor is one of the most common objections made against Rawls's method a problem either: this is the objection that it is warmed-up intuitionism.[8] The difficulties with old-style ethical intuitionism all stem from the foundational nature of the intuitions. But Rawlsian intuitions simply are not like that.

There is a serious problem, however, with reflective equilibrium when it is employed to explain the history–methodology evidential relationship, in spite of its general attractiveness. The problem is not that the method is incoherent, or that it yields what are known to be the wrong answers. The problem is one of applicability. There is no way we can use the method to justify methodological norms on the basis of historical episodes. Rawls has just one argument for using reflective equilibrium in ethics. (Such weak support for the method is itself a reason for caution.) He cites the

116

analogy between reflective equilibrium in ethics and the way he thinks the various natural sciences and linguistics are done. When there is a conflict between a scientific theory and empirical observations, sometimes the theory is revised, and at other times the theory is maintained while the observations are rejected. When to do one thing and when to do the other is taken by Rawls to be unproblematic: we should just do whatever good scientific practice says we should do.

Such a method might well be a very good method within ethics, where it is quite appropriate to appeal to the already established, or at least taken-for-granted, results of the philosophy of science. But such an appeal, quite obviously, is not open to us here. We do not yet know what the rules of good methodology are. Indeed, that is what the whole debate is about. Consequently, we could not recognize a state of reflective equilibrium even if we reached it. Progress in ethics is parasitic on progress here.

LAUDAN'S NATURALISM

Laudan has recently abandoned his PI criterion, so heavily dependent upon intuitions, in favour of a naturalistic view. A stimulus for this change of heart was Garber (1986). Garber shares with Laudan (1977) the desire to do justice to our intuitions, but claims that the cases we reflect on need not be real historical cases: 'what is important about the cases [that Laudan] cites is not their historicity, the fact that they *actually happened*, but the fact that they are a concrete embodiment of our intuitions about what is rational and what isn't' (Garber 1986:105).

In response to this, Laudan completely concedes the point. If we are to use intuitions at all then it matters not whether they are intuitions of actual cases or of fictitious ones. Garber is right, he allows; either way it is *our* intuitions that are being used for testing purposes. So, in consequence of this, Laudan says we must throw out any reliance on intuitions entirely (1986, 1987). The earlier PI criterion is then replaced (in Laudan 1984, 1986, 1987) with a kind of naturalism.

Before getting on to the new meta-methodology, let me point out some problems with the way Laudan has abandoned intuitions. First, if the problem with intuitions is that they are *ours*, then there is still Lakatos's version of intuitionism to consider. Lakatos, as we saw above, does not rely on *our* intuitions, but on the intuitions of the scientific elite. These are spread throughout history, and not just the product of here and now.

Second, Laudan over-reacts. This is a common sin in specifying a meta-criterion; people try to give both necessary and sufficient conditions for using history to choose a methodology when they should be trying only to give necessary ones. When we specify how a scientific theory should be tested by experimental evidence we do not thereby give both necessary and sufficient conditions for accepting the theory. There can be other (i.e., conceptual) considerations as well. The empirical and the conceptual are both necessary, but neither is sufficient (though together they might be).

Similarly, a methodology should do justice to history, and to give a criterion is to give necessary and sufficient conditions for it to do justice to that history. This much is necessary, but it is not sufficient for the choice of the best overall methodology. Taking account of conceptual issues here is also required. A meta-methodology does two things: (1) it does justice to history; (2) it does justice to conceptual matters. Laudan's PI criterion, his new naturalism, and Lakatos's meta-methodology based on the intuitions of the scientific elite are all attempts to give necessary and sufficient conditions for (1). They then, quite mistakenly, further claim to have provided a *complete* meta-methodology.

Garber and other more traditional a priori methodologists focus in on (2). They too are mistaken, but they realize what they are doing and usually provide some sort of argument for dismissing (1) (such as, that to countenance (1) would be to commit the naturalistic fallacy). Those like Lakatos and Laudan who mistakenly ignore (2) do so, I think, unwittingly.

Consequently, Laudan's response to Garber should not have been to attack Garber's avowal of (2), but rather to attack his disavowal of (1). In fact, that Laudan *must* embrace (2) follows almost trivially from his new naturalism. On his new view, to be discussed in detail momentarily, the way we test methodologies is the same as the way we test ordinary scientific theories. When it comes to testing the latter, Laudan stresses the crucial importance of the role of conceptual issues, so it follows that they must figure in a meta-methodology as well; we must do justice to conceptual considerations as much as to history.

In passing let me stress again what is wrong with embracing (2) and ignoring (1), as Garber does. This is the Henry Ford account of history. It leaves open the possibility that the entire history of science has been irrational, a consequence which we should all find absurd. By stressing *our* intuitions Garber makes it possible, and on occasion quite explicitly maintains, that the intuitions of past scientists have been different. If rationality is just rationality *by our lights*, then we cannot offer rational explanations of what happened in the past. We can only offer psychological and sociological accounts, but once we have done this we slip into a relativism that cannot distinguish between Newton's achievements and those of some other scientist of the time who thought the universe is a poached egg.

A third problem with Laudan's response to Garber concerns the alleged equivalence of intuitions about actual and fictitious examples. Garber claims that what is important is the intuitions themselves; whether they are about something real or something imaginary is irrelevant. Laudan concedes the point, then goes on to dismiss the use of intuitions entirely. In the light of the immediately preceding remarks, the issue takes on considerable importance.

There is a world of difference between actual and imaginary examples. There are, very likely, influences upon our intuitions which are not under our control, so there is a problem of relative contamination. Just consider the analogous situation in ordinary science. Thought-experiments, undoubtedly, play an important role in the development and appraisal of theories. But what would we think of a theory that did justice to all thought-experiments and yet had trouble with real ones?

Laudan's latest work, *Science and Values* (1984), is a little book about cognitive values (not ethical ones). It is an account of the relations among theories, methods, and goals ('aims', 'values' and 'goals' are used synonymously), and its main aim is to dissect the holistic picture of science. The author is gunning for Kuhn, but he is also, by acknowledged implication, out to debunk some of his own cherished past views. It is in this book that we find his new account of how we learn about methodology.

The stage is set in the first chapter by posing two puzzles. Laudan asks, Why is there consensus in science, and why is there dissensus? Some accounts of science, those offered by logical empiricists or by some sociologists such as Robert Merton, are capable of explaining consensus: people typically believe what has been produced by good scientific method, they would argue, since the product of that unique method is manifestly rational to accept. However, these accounts of science have the greatest difficulty in explaining the existence of prolonged disagreement. Why do people resist the 'obvious'?

On the other hand, as Laudan would have it, the accounts of Kuhn, Feyerabend, Lakatos, and the relativistically minded sociologists of knowledge can give workable accounts of dissensus in science; but they cannot explain agreement (that is, near-universal agreement) when it arises. What is needed, is an account of science which does justice to both, and this is what Laudan hopes to provide.

A common view of scientific opinion-making is that it is hierarchical; we move up the ladder from theories to methods to values. Disagreements over theories are settled by appeal to common methods. These are shared norms which, when properly applied, almost invariably decide a disagreement at the theory level. But what happens when there is a quarrel over the methods themselves? Disagreement here, according to the hierarchical view, is settled by appeal to a still higher level of common aims, goals, or values. (Laudan calls this the 'axiological' level.) Every disagreement is resolved by going up a level in the hierarchy. But at the level of values we reach the end: there is nowhere to go to settle a dispute which is at this level. For most empiricists, values are just not cognitive; one cannot claim to be rationally adopting or rationally abandoning various different scientific goals. On an empiricist epistemology they are mere preferences, which the rest of us can like or hate; but that is all there is to it. These sentiments are embodied in the positivist doctrine that the verifiability principle is not itself meaningful, and in Popper's falsificationism, which is a conventionally adopted methodology and is not itself falsifiable (at least as it is found in *The Logic of Scientific Discovery*; his later view in *Objective Knowledge* portrays his methodology as an objective truth about the 'third world').

The general structure of the hierarchy scheme as Laudan views it is as follows:

Level of Disagreement	*Level of Resolution*
Factual	Methodological
Methodological	Aims and values
Aims and values	None

(Laudan 1984:27)

120

But this hierarchical account leaves us in limbo; we now seem unable to explain agreement. Are values really not arrived at rationally? Of course they are, and Laudan quite rightly claims that we can indeed rationally evaluate cognitive aims in a number of respects. For instance, one way is on the basis of consistency (though non-cognitivists require that too). More interesting is a type of criticism of goals that Laudan calls 'utopianism'. We can sometimes show that a particular goal is unattainable. One might have a goal of building a perpetual-motion machine; yet having such a goal is surely not beyond criticism. On the other hand, some goals might be attainable, but not recognizable if stumbled upon. Truth and simplicity, for example, would not be manifestly obvious even if we should attain them; so, one might conclude, they should not be made our goals. Laudan argues quite explicitly against making truth an aim of science in his discussion of realism. However, one might take the view, in opposition to Laudan, that truth can be made a goal, and then argue that the presence of some discernible property P is a sign of truth. Consequently, seeking theories with property P becomes a sub-goal, or perhaps a method of attaining that goal. I realize, of course, that this 'might' is a mighty big might, since we've been arguing for two-and-a-half thousand years about what the signs of truth look like.

The final grounds offered by Laudan for rationally evaluating aims concern the tension that often arises between explicit aims on the one hand and actual practices on the other. Close scrutiny of actual scientific theorizing often suggests that there are quite different methods at work than the ones dictated by the level of goals. The tension between what is practised and what is preached is nicely illustrated by the case of eighteenth-century aether theories. Hartley and LaSage fell foul of the prevailing empiricism by hypothesizing a hidden mechanism to account for various phenomena. Scientists of the day had either to abandon several cherished theories which posited theoretical entitites, or else abandon the methodological rule which enjoined them not to go beyond the observable data.

Though Laudan is attacking the hierarchical account, he is careful about whom he makes common cause with. One line of thinking which also rejects this account is based on underdetermination considerations. It runs as follows: there are infinitely many theories which do justice to a given methodology, so no dispute over theories can be resolved by going up to a level and appealing to methodological considerations. Laudan completely rejects this (Laudan 1984:28f.). Underdetermination, he rightly argues, does not undermine this hierarchical view of theories/methods/aims. The problem is laid to rest by the fact that there simply aren't infinitely many theories to choose from. In a logical sense there are, of course, but from any practical point of view, there are only a very few live options on the table. And here, using the existing methods, we can usually decide between the actual rivals which are at hand. (This sort of consideration was used in chapter three.)

This seems to be exactly the right response to any who would attack the rationality of theory – choice by saying we have infinitely many equally adequate choices. But after giving this argument, Laudan goes on to discuss methodological disagreements. Instead of using similar considerations, he raises the possibility of an underdetermination at this level (1984, p.35). Why, he asks, couldn't there be more than one way to achieve one's cognitive goals? Two scientists with the same axiology, that is, with the same goals, may still have different but equally good ways of realizing them. For reasons which are not clear, Laudan thinks a plurality of methods is desirable while unique choices at the theory level are called for. But why, the reader is bound to wonder, does Laudan not use the same argument as before, i.e., that the number of actual methodological alternatives is finite and they are most likely not to be of equal merit? He seems to shift his ground rather arbitrarily to suit his immediate purposes. When he wants to have scientific consensus he has no underdetermination of theory-choice, but when he wants prolonged disagreements in science, methodological underdetermination is readily at hand.

Perhaps it is because he has something else up his sleeve. Laudan doesn't want methodological disagreements to be always settled by going up a level in the hierarchy. Instead, he claims, there are empirical resources for helping to decide methodological norms. For example, the discovery of placebo effects at the theory level in psychology led to significant methodological changes, so that now control groups, blind and even double-blind tests are considered essential practice in a variety of situations. Examples such as this break the hierarchy ideal, for here a dispute was resolved not by going up, but by going down a level; a theory overruled a method.

The standard hierarchy model, as I mentioned above, stops at the level of aims. Aims and goals are non-cognitive values, and there are no rational choices to be made here according to typical empiricists. In contrast, Laudan claims that 'methodological norms and rules assert empirically testable relations between ends and means . . . [and so,] construed of course as conditional imperatives (conditional relative to a set of aims), should form the core of a naturalist theory of scientific knowledge' (Laudan 1984:40). It may be true that many hierarchical accounts stop at values, and this is undoubtedly true for positivistic versions, but not everyone who subscribes to the hierarchy is a non-cognitivist about values. One might hold that values are open to rational evaluation, but that the ground of this appraisal lies elsewhere than below. William Whewell thought the aim of science was to discover the works of God. This in turn suggested a methodology having to do with the fact that the human mind and God's creation were to one another as hand to glove (since whenever God wants us to do something he gives us the necessary tools). But where did this aim itself come from? It came from Whewell's theology, and Whewell would have been prepared to present arguments for it. His was a hierarchical model of science, but it did not stop at aims and values. If it stopped at all, it would have done so at religious faith, but it may not have stopped at all.

There is an ongoing ambiguity throughout Laudan's book between methods and values. Consider the following discussion:

> if it comes to a choice between Kepler's laws and Newton's
> planetary astronomy . . . and if our primary standard is, say,
> scope or generality of application . . . then our preference is once
> again dictated by our values. . . . Under such circumstances the
> rule, 'prefer theories of greater generality', gives unequivocal
> advice.
>
> (Laudan 1984:32)

Laudan calls it a value, but the statement 'prefer theories of greater generality' seems much more like a methodological rule. In another place he remarks that the desire for 'theories that are true, general, simple, and explanatory' are cognitive values. But surely this would be much better viewed as having 'truth' as the single value, and as having simplicity, generality, and explanatory power function as *signs* of truth. Once we have them as signs that we are on the road to our goals, we can then make methodological rules out of them, i.e., 'accept theories with greater simplicity', and so on. (Of course, whether they are good rules is debatable.)

On the other hand, if we want to keep several of these as values, then very likely it will have to be as pragmatic values, not cognitive ones. Van Fraassen's attack on realism fails, I think, but it is not entirely unsuccessful. This much at least is right: if science does not aim for truth, but rather for empirical adequacy, then such theoretical virtues as simplicity and explanatory power are part of the pragmatics of theory choice (see van Fraasen 1980). Economic considerations provide some of the best illustrations of pragmatic properties (though they are not ones van Fraassen discusses). A theory which costs only half as much to test as another has some sort of virtue, but not a cognitive one. Economic values can even influence methodological principles, for example, 'All other things being equal, test the theory which is cheaper to test first' (see Rescher 1976). But the role of such methodological rules is not cognitive.

Less ambiguous, but just as problematic, are some of Laudan's other choices for illustration purposes. In arguing that goals can themselves be rationally criticized, Laudan cites as an instance the aim, 'to travel at velocities higher than the speed of light' (1984:51). Laudan is undoubtedly right in thinking such an aim unreasonable. But it is also a technological or instrumental goal; it is certainly not a cognitive one. The plausibility of Laudan's case is partly built on such debatable examples.

As I mentioned above, Laudan uses the introduction of the method of hypotheses by Hartley and LaSage in the eighteenth century to illustrate some of his views. LaSage had a theory of gravitation based on the impact of ultramundane corpuscles. Laudan says that this theory was in conflict with the 'aims and goals of science' of the time. Surely not; the aim then was truth, just as it is now; the quarrel was over how to get there. The introduction of hypothetico-deductivism was a change in method, not a change in aims.

It should be quite clear by now, whether we are convinced by his arguments or not, what Laudan wants to do. He wishes to turn the hierarchy into a circle or a network. Instead of the hierarchy, Laudan offers in its place the triad, with each element interacting with the other two (1984:63):

124

methods

theories ⟷ aims

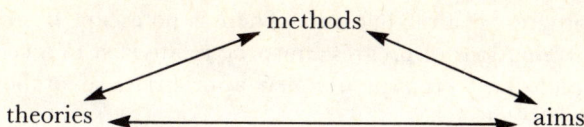

According to Kuhn, who is Laudan's main foil in this book, paradigm change is a change of the whole network; old theories, methods, and goals are all abandoned simultaneously in favour of new theories, methods, and goals. That is, one triad (which I will denote $\{T,M,V\}$) gives way to another $\{T',M',V'\}$, which differs in every element. A paradigm is an 'inextricable mix' of these three features; to change one is to change all. The disturbing aspect of Kuhn's view is that paradigm change seems non-rational. The regular charge against Kuhn, a charge which Laudan endorses, is that if the very techniques of assessment are changing as well as the theories themselves, then there is nothing neutral and non-question-begging left over with which to assess the paradigm change. To this Laudan has a very simple and plausible alternative. 'Provided theory change occurs one level at a time, there is ample scope for regarding it as a thoroughly reasoned process' (Laudan 1984:75). So, instead of transitions of the form:

$$\{T,M,V\} \rightarrow \{T',M',V'\} \rightarrow \{T'',M'',V''\}$$

which is the Kuhnian picture of things, we have on Laudan's view something much more piecemeal, perhaps like this:

$$\{T,M,V\} \rightarrow \{T,M',V\} \rightarrow \{T',M',V\} \rightarrow \{T'',M',V\} \rightarrow \{T'',M',V'\}$$

Though allowing that it would take a lot of historical work to show this, Laudan claims that history will vindicate his view of scientific change over Kuhn's; that is, the great bulk of what we think of as the rational history of science will fit Laudan's pattern.

Taking the development of the triad of aims, methods, and theories to characterize the evolution of science, Laudan then goes on to talk about progress. At first there seems to be a paradox. How can science make progress if the aims of science keep changing? Progress, after all, must be progress toward some goal, but no goal will stand still; the aims are evolving. Whigs take their own aims as uniquely desirable and then discuss the history of science in terms of progress toward them. This, of course, gives us a mistaken view of the historical development of science. As for Laudan, rather than resolve the paradox, he embraces it. The only kind of progress is a relative one. 'There is no escape from the fact that determinations of progress must be relativized to a certain set of ends, and that there is no uniquely appropriate set of those ends' (Laudan 1984:66).

However, this relativism is to some extent undermined by a few simple observations. To start with, inside any given triad we can, as Laudan argues, recognize tensions; we want to overcome those tensions and we usually have ways of doing so. Does it not follow from this, trivially in fact, that we must have a transhistorical goal? Why else would we tinker with any triad? And does it not further follow that we have a transhistorical method for doing so? If we do have such a goal and a way of achieving it then there is indeed a transhistorical sense of scientific progress; it is progress brought about by harmonizing the {T,M,V} triad. I don't want to make much of this, since it is achieved far too easily. Instead let me pick up on a point alluded to above, i.e., that Laudan was not at all convincing about changing aims in science. Not all of his examples were instances of cognitive aims; some were either aims of another sort or disguised methods. Even the Whewell example of having the aim of understanding God's creation can be seen as not really an aim of science, but a moral injuction to do science.

The actual room for manoeuvre in shifting scientific goals strikes me as rather small. The debate between realists and anti-realists may circumscribe it. 'Seek truth' or 'Seek empirical adequacy' pretty well exhaust the history of discernible scientific goals. Attacks on the aims of science which attempt to show that those aims are incoherent or hopelessly utopian tend to be of the philosophical sort which are typically ignored by scientists. We cannot prove that we're not fooled by Descartes's evil genius, but such principled scepticism plays no role *in* science. If we came to accept such a challenge to the present aims of science we wouldn't keep our theories and method and merely change our aims. We would give up science entirely. Philosophy touches science in several ways, but neither philosophy nor anything else has much impact on the cognitive aims of science. In consequence, the influence of theories or methods on cognitive values may be vanishingly small. If so, then a much better picture of the triad would then look like this:

eternal values

theories ⟵————————————————⟶ methods
(low level)

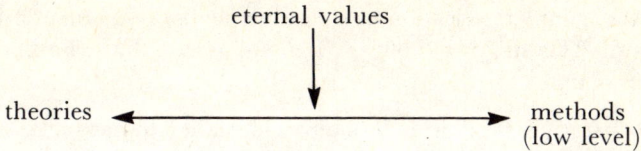

On the other hand, I want to grant just about everything to Laudan concerning the theory–method relation; namely, that it is a two-way interaction. This in itself is enough to break one aspect of the overdone holistic picture, for we can indeed have methods held static during a change in theory, and vice versa.

When we talk as Laudan does so freely about methods evolving, however, we must be very careful. Liberal methodologists are often confused on this. Even *Progress and its Problems*, written when Laudan was in his 'holistic' phase, which he is now repudiating, would occasionally run global methodology together with local methodology. The latter comprises norms such as the requirement of 'blind testing', or the mechanical philosophers' injunction to 'posit only mechanical interactions'. Laudan quite rightly notes that these low-level techniques were adopted and rejected at various points in the history of science.

On the other hand, Laudan also had a different level of methodological norm functioning in his earlier book. This is the level at which we choose among rival research traditions, (Research traditions are global units consisting of a sequence of substantive theories about the world and also of 'methodological *do*s and *don't*s'.) In his earlier work Laudan posits rules for choosing among rival research traditions: Accept the one with the greatest momentary adequacy; pursue the one which is solving problems at the fastest rate. These global or high-level rules are absolutely eternal (if true at all) whereas local or low-level norms, such as the prescriptions of the mechanical philosphers, come and go with various research traditions. The distinction between global methodology, which is eternal, and the lower-level, paradigm-bound methods, which change, is surely a distinction on the right track.

Laudan has moved away from his earlier position to such an extent that he now declares 'the requirement that methodology or epistemology must exhibit past science as rational is thoroughly wrong-headed' (Laudan 1987:20). Moving away from his earlier position involving the pre-analytic intuitions is understandable, but why the complete rejection of the principle that we must do justice to history? Calling it a 'howler', Laudan asserts that what has gone wrong is that

> the historicists' meta-methodology has failed to reckon with the fact that *both* the aims *and* the background beliefs of scientists vary from agent to agent, and that this is particularly so when one is talking about scientific epochs very different from our own. If the aims of scientists have changed through time in significant respects, we cannot reasonably expect *our* methods – geared as they are to the realization of *our* ends – to entail anything whatever about the rationality or irrationality of agents with quite different aims. Whatever else rationality is, it is agent- and context-specific. When we say that an agent acted rationally, we are asserting minimally that he acted in ways which he believed would promote his ends.
>
> (Laudan 1987:21)

But all of this is beside the point, as 'methodology' is used far too loosely. And, besides, what cranks don't think they are promoting their own ends?

The problem with holism that Laudan rightly attacks is that it links theories and methods too tightly together. We can jettison that link without also tossing out the low-level/high-level distinction which did valuable work and which ought to stay. It is this eternal methodology which tries to achieve our eternal goal.

THE THEORY-LADENNESS OF HISTORIOGRAPHY

I shall now take up the topic of the nature of 'observation', especially as it pertains to historical investigation. The notion is unclear and problematic, yet important, because the relation between observing, on the one hand, and the great body of our beliefs, desires, and expectations on the other, is far from straightforward.

The question is this: can something akin to neutral observations be made of historical events, or are such perceptions conditioned, influenced, and even prejudiced by prior beliefs, hopes, and anticipations? In short: are historical observations theory-laden?

What does a historian look at anyway? Archive materials, for one thing: books, letters, and notes. Implements and artefacts for another: Galileo's telescope, for example. The historian will also, on occasion, try to reproduce certain experiments and results: Newton's rings, Fresnel's diffraction experiments, Faraday's magnetic induction work, and so on. In such examples as these, is it the case that the historian's perceptions are theory-laden? Will rival historians have different perceptions of what went on? In order to answer this we need first to make some distinctions.

Consider, to begin with, the case of two rival historians who are both looking at, say, a magnet, or Newton's rings, or a diffraction pattern. Do they see different things? Not if they have the same beliefs and expectations about the world. They may differ in their accounts of Gilbert on the magnet, but they agree on what they take to be the factual truth of the matter. The same goes for perceiving Newton's rings or a diffraction pattern. If both historians accept *contemporary* optics then they will both see the same thing. Using Hanson's famous example: if Kepler and Tycho had held the same planetary theory then they would have seen the same sun. Thus, contemporary historians may not see what historical personages saw, but each will see the same thing as other contemporary historians since, as a rule, they will share the same theories about the world, namely our best contemporary theories. Are such observations neutral? In one sense, no. They are theory-laden, at least if Kuhn, Hanson, Feyerabend, and others are right, and for present purposes let's assume they are. But in another sense these observations are neutral, because they are the same for all historians. So, rival historians need not be in conflict on this score.

Second, historians often try to put themselves into the historical figure's frame of mind. They try to 'get into Newton's head' and see things the way Newton saw them. If two historians – we imagine them still to be rivals – were both successful in getting to see things the way Newton saw them, then, consequently, each historian would be seeing things the same way the other sees them. When they put on their seventeenth-century Newtonian-mechanics thinking caps they see the world alike, and moreover, they see it just as Newton did. Of course such perceptions are theory-laden. And the theory is a different one from contemporary theories, so the historians see the world differently with their Newtonian thinking caps on from the way they see it with their contemporary-science thinking caps on. But the important thing is this: any historian who sees things the way Newton saw them will see things the same way as any other hitorian who sees things the way Newton saw them. Once again we have neutrality of observation; not absolute neutrality, but neutrality in the sense of inter-subjective agreement about what is being seen.

Third, there are still other possibilities. A notebook entry, for instance, might be smudged. One historian expecting the author to have written x may well see x, whereas another historian with different expectations might see y. But this will be a relatively rare event. The debates historians have among themselves hardly ever turn on this sort of thing. Doubtless, it does sometimes occur and we should note its existence; however, it is not very significant. If we are to take seriously the thesis that historical observations are theory-laden, it must be in some other sense of the word. A historian's sensory perceptions will be of little concern to us, since the perceptions of rival historians will be the same.

Let us try a different tack. Historiographies of science are explanatory, but the way they explain is sometimes in one respect quite different from the way an explanation functions in the physical sciences. Explanations for human actions often appeal to reasons and rationality considerations. (This, of course, is what was argued for above.) So, for instance, we might ask why Galileo accepted Copernicanism when he did. An explanation might go like this: Galileo had considerable telescopic evidence for Copernicanism, and one should always accept a theory with such support. In other words, we explain the event by pointing out the rationality of the action. The evidence is the cause of the belief. We show that, in the circumstances, the action performed was the rational thing to do. (And if we cannot give a rational explanation then we try for some other, usually of a psycho-social nature.) Considerations of rationality must appeal to some methodology or other. This is the point where norms and values come into play.

There is a trivial sense in which values get injected into the history of science. A historian chooses, perhaps on the basis of taste or perhaps because of some social interest he or she is promoting, whether to write a book on Copernicus or on Darwin. This is no different from the same trivial injection of values into the natural sciences. (But it is not always trivial, as we shall see in the final chapter.) A scientist chooses, on the basis of interest and temperament, to pursue astrophysics rather than microbiology.

This is not sufficient for getting values into the very content of history. A much more interesting injection of values occurs when the historian sorts through all of the available material and picks out what is 'important'. To do this the historian must have some idea as to what it is that makes for good science. This usually involves such things as choosing to include the results of experiments and deciding to note what background assumptions were held by the scientist in question. What the scientist had for breakfast or where he or she took summer vacations is usually omitted.

We cannot explain actions in terms of their rationality without presupposing some account of rationality or other. That is, we must drag in some normative philosophy of science. And with it come those special concepts and categories which are often unique to particular philosophies of science. Consider, for example, concepts like 'crucial experiment', 'falsified theory', 'confirmation function', 'research programme', 'positive heuristic', 'paradigm'. Clearly, we do not observe, as a neutral given, the crucialness of a particular experiment. Also a research programme is a sequence of theories, an ordered set; but a set is an abstract entity which can no more be seen than humankind or the number seven.

Yet some historians claim they can see crucial experiments while others are blind to them; some see evidence and method where others see only irrelevant data and madness. There seems but one inference to make: these historical observations are theory-laden. Seeing crucial experiments, or seeing research programmes, requires seeing events through the eyes of some theory, a normative theory, a theory about how science ought to be practised.

We believe there are electrons not because we can see them directly, but because we accept the particular atomic theory which says they exist. Analogously, we believe there are crucial experiments (if we do so believe) because we accept the particular normative theory which says they can occur and which tells us how to recognize them when they do.

History is theory-laden in the senses I mentioned at the outset of this section; but these ways of being theory-laden are of little significance to us. The most important sense, for our purposes here, is that a historiography must employ concepts from some normative theory. A thorough history of science is explanatory, and the correct explanation for many cognitive decisions is in terms of their rationality. The only way to give such an explanation is by invoking methodological concepts. This is the significant sense in which history is theory-laden; it is laden with normative theory. When rival historians see different things it is usually because they are seeing the past through the eyes of different normative philosophies of science. Of course, this sort of perception is not sensory perception, but it is a kind of perception none the less, and it *is* theory-laden.

These considerations set the stage for my own view of the history–methodology evidential relationship which now follows.

Chapter Six

HOW TO BE
AN ANTHROPOLOGIST
OF SCIENCE

In an earlier chapter I accepted the anthropological analogy: we should approach science just as we would approach any exotic culture. The difference between rationalists and sociologists of science is really just a difference in their respective theories as to how science works. What we have are rival accounts of the causal structure of belief in scientific society; sociologists take belief structure to be the result (usually) of social forces, and rationalists take it to be due (usually) to available evidence and good reasons. If we want to talk this way, both accounts are perfectly scientific; they are simply rival anthropological theories.

Bloor's science of science, the finitism of Barnes, and the laboratory accounts of Collins, and of Latour and Woolgar, are views about how science must work. So far I have argued (in chapters two, three, and four) that the case they make for social causation has not been successful. This leaves the rationalist alternative. In the last chapter some rationalist views on the role of case studies were criticized. There I argued that case studies do indeed constitute an empirical study of scientific society. I now aim to show how these case studies provide evidence for any theoretical view about how the society of science works. By saying how empirical evidence supports or refutes a view, I am in effect saying how to be an anthropologist of science.

I'll begin by picking up the strand from the last chapter. There it was argued that an evidential relation between historical facts and methodological norms exists, but that accounts so far proposed of that relation are faulty in some way or other. A new account is needed.

I claim that the following criterion for justifying a normative methodology is the best characterization of the evidential relation:

> *(R)* That methodology is best which (all other things being equal) makes
> its theoretical reconstructions and normative reconstructions coincide
> for the greatest number of episodes in the history of science, while
> cohering with relevant independent sociological theories.

Just as theories of scientific method provide criteria by which rival
scientific theories are to be evaluated, so *R* provides a *meta-criterion*,
by which rival methodologies are to be evaluated.[1] As I stressed
above when discussing Laudan's naturalism, a methodology should
do two things: it should do justice to history, and it should also do
justice to conceptual concerns. *R* is only trying to satisfy the first of
these; everything else is covered by the 'all other things being
equal' clause. I don't mean to downplay conceptual or a priori
issues in normative methodology; it is just that I am concerned
here with how the 'empirical' aspect of things works.

Some explanations of criterion *R* are in order here. The term
'rational reconstruction' has been used by several philosophers,
especially Lakatos, in a number of distinct and equivocal ways; so
I shall avoid it in favour of two new terms, 'theoretical
reconstruction' and 'normative reconstruction'. In the previous
section, I noted that historical observations, and hence historical
descriptions, are theory-laden. It is methodological theory that
they are laden with. A *theoretical reconstruction* is supposed to capture
this notion. It is a description of a particular historical episode
using the concepts of some methodology. For example, if we
describe a historical episode using Popperian methodological
concepts, then the account will be in terms of 'crucial experiments',
'basic statements', 'falsified theories' and so on. On the other hand,
if we were to describe the same episode using Lakatos's so-called
'methodology of scientific research programmes', then the history
would be written employing such terms as 'research programmes',
'heuristics', 'hard core', 'progressive', 'degenerating'.

These terms, 'crucial experiment', 'heuristic', and so on, are
theoretical terms. The concepts come from a methodological
theory. Any written historiography of science must employ such
normative concepts if it is to be explanatory at all. It is in principle
impossible to write a history of science without utilizing such
notions, that is, without using concepts which owe their being to
some normative philosophy of science (at least, if it is a rationalist
history of science; a sociological history will invoke different

theoretical entities, such as 'professional' or 'class interests' to explain events). When a history does describe some episode using the conceptual apparatus of some methodology it is to be called a 'theoretical reconstruction'.

The reason why a historiography must invoke methodological concepts if it is to be an explanatory account is very simple, but it should be stressed. In order to explain an event we cite the cause. In the case of a cognitive decision, this amounts to citing a reason – at least, if the decision was a rational one. Thus, the reason why someone switched theories could, for example, be the occurrence of a crucial experiment. So, to explain the event is to cite the cause, which is to cite a reason, which is to invoke a methodological concept, i.e., the crucial experiment.

Whereas a theoretical reconstruction is an attempted description of actual history, a *normative reconstruction* is quite another thing. Like a theoretical reconstruction, it employs the conceptual apparatus of some methodology or other: but a normative reconstruction does not attempt to say how history *actually* went. Rather, it declares how history *ought* to have gone according to that methodology. The initial starting-point for the theoretical and the normative reconstructions will be the same, but from that point forward they may or may not diverge.

A simple example will illustrate these two notions. We might have the following (very simplified) theoretical reconstruction using, say, Popperian methodology: Jones boldly conjectured a theory T and submitted it to a crucial test. T was falsified. Jones continued to maintain T and looked for ways to confirm it. On the other hand, using the same Popperian methodology, the normative reconstruction of the historical episode would run: Jones boldly conjectured a theory T. He submitted it to a crucial test. T was falsified. Jones rejected T

The same historical episode might be theoretically reconstructed as follows, this time using, say, Lakatos's methodology of scientific research programmes: Jones conjectured a theory T which was part of a research programme RP. T made a number of empirical predictions, some of which succeeded and others of which failed. Jones did not abandon RP, but made slight modifications, which amounted to changes in the protective belt, leaving the hard core of RP intact. Now we switch to the normative reconstruction of this episode, still using Lakatos's methodology: the account, let us suppose, runs exactly the same as the theoretical reconstruction. That is, Jones conjectured a theory T which was part of a research programme RP. Some of T's predictions panned out while others didn't. Jones did not abandon RP but made modifications in it which amounted to a new theory, leaving the hard core of RP intact and modifying only the protective belt.

What I have called a theoretical reconstruction is an attempt, albeit highly theory-laden, to describe how history *actually* happened. And what I have called a normative reconstruction is an account of how history *ought* to have gone according to that methodology. My rule R says that it is a plus for a methodology if the theoretical and the normative reconstructions coincide in any particular episode. In the Jones case, Popperian methodology does not lead to a coincidence between the theoretical and the normative reconstructions, but Lakatosian methodology does. Thus, the Jones episode is evidence *for* Lakatosian methodology and evidence *against* Popperian methodology.

If all of this seems too elaborate, a simple example from physics may help it appear familiar. Suppose we see a great mishmash of streaks in a cloud-chamber photograph and interpret it as the decay of a neutron into an anti-proton, a positron (or anti-electron), and a neutrino: $n \rightarrow \bar{p} + \bar{e} + v_e$. This is a highly theory-laden observation. It corresponds to what I have called a theoretical reconstruction of historical events. A particular theory of elementary particles may say that if a sequence starts with the decay of a neutron it must continue as follows: $n \rightarrow p + e + \bar{v}_e$. This is a prediction; it corresponds to my normative reconstructions. (The theory rules out processes such as the one described first, since they violate conservation of baryon number.)

R says we should think more highly of a methodology when theoretical and normative descriptions coincide, just as we think more highly of a theory when the description of actual events (theory-laden though that description is) and the prediction of those same events coincide.

A great many historically oriented, rationalist-minded philosophers of science have been at considerable pains to present careful case studies. But what role did these case studies play? Clearly, they are intended to do more than merely illustrate some new a priori philosophy of science; they are supposed to be some sort of evidence. But how? It has never been clear, except in those few cases criticized in the last chapter. R now very straightforwardly says just what it is that the case study does; it says just how the case study serves its evidentially supporting role.

Many will be tempted to lambaste R. They will see it as implying that the best methodology is the one which makes *everything* rational, and we do not have to be cynics about the history of science to believe that that is an absurd consequence. But why do we doubt that the entire history of science is rational? Surely, it is because we have background beliefs about typical human behaviour. We know that on occasion people make mistakes, cheat to promote their own interests, and often enough are just plain silly. These are perfectly reasonable beliefs, not the result of some jaded misanthropy. They are covered, however, by the final clause in R. Let me elaborate.

Recall that when I discussed Lakatos, one of the objections raised against him was that on his account the best methodology might be the one which made everything rational (at least everything the scientific elite thought was rational). This will not occur on my account. The proviso in R prohibits this sort of absurdity. The coherence demanded by R tells us to maximize rationality as far as we can while still cohering with the best psycho-social theories (which in effect tell us to what degree scientists can be expected to be irrational). There is no question of R making every episode in the history of science rational. A methodology which did would violate R.

At this juncture I should give examples of the sorts of theory, especially sociological theory, that a methodology must cohere with. I'll turn to one of the most prominent sociologists of science, Robert Merton, for illustrative examples. Whether his theories are true or false does not matter here, since my aim at this point is merely to illustrate R. According to Merton (1973), the scientific ethos is characterized by the following properties: (1) Universalism: the evidence is open to all; there are no privileged observers. (2) Communism: knowledge is collectively arrived at and owned by all. (3) Disinterestedness: we approach nature without prior wishes that it be one way or another. (4) Organized scepticism: nothing is immune from doubt. These are the values, says Merton, that most scientists explicitly and implicitly hold. Moreover, these are the values that the scientific community as a whole promotes.

There is a conflict, says Merton, between the reward system of science and these values which characterize the scientific ethos. The reward system of science is based almost entirely on peer recognition, specifically, recognition for originality. Scientists are obliged to advance knowledge and the only way to do that is by being original. The pay-off is in the form of having the contribution acknowledged. (Often this takes the form of eponymy: for example, Newton's Laws, Planck's constant, Mendelian genetics, the Freudian slip, Maxwell's equations, and so on.) Merton thinks that it is not so much that egotistical people enter science as a way to become recognized; rather, the institution makes them desire it. The institution of science makes originality and the recognition that goes with it an imperative. This is the only way, says Merton, that we can explain why such ordinarily shy and self-effacing people as, say, Henry Cavendish get caught up in disputes over priority of discovery. The desire for recognition does not stem from an innate egoism, but must instead stem from institutional sources. Thus, according to Merton, we have an inevitable conflict between the reward system of science and the scientific ethos.

There are a number of different responses that scientists make to these conflicting demands. One is to do brilliant, original work which is recognized by the community. This, of course, is not a path open to everyone. Other responses include fraud, hoax, plagiarism, the construction of anagrams, and, most worrisome of all, the prudential cooking of data. The reason this last is so worrisome is that it may be hard to separate it from honest but theory-laden observation.[2] There is, of course, a difference between theory-laden and ambition-laden observations, but it isn't a difference that is easy to detect.

Frauds, hoaxes, data cooking, and so on, are forms of deviant behaviour which result in part from the conflicting values of the scientific institution. Merton proposes the following general law: 'Any social institution which gives emphasis to aspirations for all, which cannot be realised by many, exerts a pressure for deviant behavior' (Merton 1973:321). Merton has also detected what he calls the Matthew Effect (1973:457ff.). The name comes from St Matthew: 'For unto every one that hath shall be given, and he shall have abundance; but from him that hath not shall be taken away even that which he hath.' Put bluntly: the rich get richer and the poor get poorer.

The Matthew Effect can manifest itself in many ways in science.

For instance, credit in the case of simultaneous discovery goes disproportionately to the more famous of the co-discoverers. And a new contribution to the scientific literature will be much more visible if it is made by a scientist who is already famous. When it comes to the allocation of resources, Merton notes that in 1962, 38 per cent of US federal support for science went to only ten institutions; the situation has not changed significantly since then.

There is a recent nice illustration of the Matthew Effect in Cesi and Peters (1980). These two sociologists selected a number of articles which had already been accepted for publication in various scientific journals, but had not yet appeared. They changed the titles, the authors' names and institutional affiliations so that fame and prestige were considerably reduced. The articles were resubmitted to the same journals. Three-quarters of them were rejected on the second submission by the journals which had previously accepted them. They concluded that being a famous author or being affiliated with a prominent institution has something of a 'halo effect'.

There is a phenomenon noted by social psychologists (and stressed by Barnes (1985:79)) concerning peer pressure and authority. When everyone else deliberately exaggerated the length of an object, the test subject tended to agree with the judgements of the peer group.

Sexism in science is another recognized sociological phenomenon. As reported by James Watson in *The Double Helix* (1968), he and Francis Crick largely ignored the X-ray diffraction work of Rosalind Franklin because she was a woman.

There are also psychological constraints on acceptable normative theories. For instance, only self-conscious logicians and statisticians could actually make explicit use of Bayes's theorem in decision-making. No one prior to Bayes's discovery could possibly have thought (that is, computed) in accordance with the theorem. We can say, almost a priori, that it is not the methodology that was actually used by Galileo or Newton. It is not enough that a methodology reconstruct history so that the episodes in question come out rational; it must also be the case that the methodology employed in the theoretical and normative reconstructions is a methodology which, psychologically speaking, could have been actually used. We are trying to find the one which actually was used, and Bayesianism, I suspect, is far too complicated for anyone to have actually calculated in accord with, at least prior to the mid-eighteenth century. (Remember that one of the things we want a methodology for is to explain why people believed what they did. So even if a Bayesian analysis of an episode made it rational, still that could not have been the *cause* of the decision.)

Psychological constraints, sexism, the Matthew Effect, and the pressures for deviant behaviour brought about by the reward system of science are typical sociological phenomena at work in the scientific community. These are the sort of things that a methodology must cohere with according to *R*. Consequently, a reconstructing methodology will likely diverge in its theoretical and normative reconstructions when it considers the behaviour of Watson and Crick with respect to their appraisal of the work of Rosalind Franklin. Should the correct methodology diverge in its theoretical and normative reconstructions of this episode then we can say that it truly was irrational of Crick and Watson to behave as they did.

One thing that must be stressed is this: we can establish such things as sexism in science and the existence of a Matthew effect without begging any normative questions. I do not say in advance that ignoring Franklin's work because she was a woman was irrational, only that it did happen. And I do not say in advance that the reconstructing methodology must diverge in its theoretical and normative reconstructions on this point, but only that it must cohere with the best psychological and sociological theories available. It may well turn out – though I certainly doubt it – that sexism is rational. But then note that by *R*, if it was rational to ignore Franklin because she was a woman, then it was *ir*rational to pay attention to Marie Curie.

There are a number of things that can be said in favour of the criterion *R*. I shall now outline a few of them.

WHY RULE *R*?

First, *R* captures the spirit of the sentiment expressed at the outset; i.e., the history of science must in some significant sense be considered rational. *R*, by trying to maximize rationality in the history of science, captures this spirit and so it has at least prima facie plausibility.

Second, *R* overcomes each of the problems of Laudan's PI account. Recall some of the difficulties with his criterion. They were, for example: (1) The PI criterion is non-comparative. (2) It is foundationalist; and not only is this doctrine in widespread disrepute but it is in conflict with Laudan's own general methodology, just as (1) is. (3) It relies on intuitions which can be understood either foundationally or as common beliefs. Neither of these interpretations of 'pre-analytic intuitions' yield a satisfactory criterion. (4) Most of history plays no role in the testing process. And lastly, (5) the criterion is motivated by the pseudo-problem of circularity.

Third, as I will show in the next section, R avoids any sort of circularity in the use of history for testing rival methodologies. The circulatory problem is the alleged difficulty that results from using a methodology to interpret history and then using that very same history to test the methodology. This has been a great problem, not only for Laudan, but for many rationalist-minded philosophers who are sympathetic to the claim that history is evidentially important. They have thought, incorrectly, that there must be some sort of difficulty with such a procedure: that it would be impossible to get a fair test. They fear the method is question-begging or self-authenticating. But I will show that this need not be so; R does not lead to the kind of circular testing procedure which was claimed to be inevitable. In fact, R completely obviates the problem.

Fourth, R is formally similar to the principle of charity. This is a methodological principle of linguistics which says that when translating the sentences of a radically different cultural group into our own, we should, as far as possible, translate their sentences into true sentences. The principle was first enunciated in Wilson (1959); it is now advocated and employed by many, including Quine, who writes:

> The maxim of translation underlying all this is that assertions startlingly false on the face of them are likely to turn on hidden differences of language. This maxim is strong enough in all of us to swerve us even from the homophonic method that is so fundamental to the very acquisition and use of one's mother tongue.
>
> The common sense behind the maxim is that one's interlocutor's sillines, beyond a certain point, is less likely than bad translation . . .
>
> (Quine 1960:59)

Donald Davidson, as we saw in a previous chapter, also uses the principle:

> What matters is this: if all we know is what sentences a speaker holds true, and we cannot assume that his language is our own, then we cannot take even a first step towards interpretation without knowing or assuming a great deal about the speaker's beliefs. Since knowledge of beliefs comes only with the ability to interpret words, the only possibility at the start is to assume general agreement on beliefs. We get a first approximation to a finished theory by assigning to sentences of a speaker conditions of truth that actually obtain (in our opinion) just when the speaker holds those sentences true.
>
> (Davidson 1973:18f.)

As a rule of thumb the principle of charity has a good deal of plausibility to it. It does have some shortcomings, but what I want here is to play on the initial plausibility of the principle and its similarity to R: as charity tries to maximize truth, so R tries to maximize rationality.

On the other hand, there is an important difference. Charity says to interpret *others* as believing what *we* believe. This is how it maximizes truth. R, to put it simply, tells us to make the greatest number of scientists come out rational whoever they may be. In consequence, *we* may have to radically overhaul what we take rationality to be. R, unlike charity, is not an imperialistic principle.[3]

Let me elaborate on this a bit. Charity tells us to maximize the truth of the beliefs of others. This principle has been rightly criticized on a number of scores. In its stead, a principle of humanity has been suggested (Grandy 1973), which has been looked upon with increasing favour (for example, in Lukes 1982). Humanity says that we should make the beliefs of others as intelligible to us as we can. This does not mean making them true as far as possible, but rather making it clear to us that, in the context, the beliefs are quite sensible.

At first glance, it looks as if my R is just the principle of humanity by another name. Assuming a belief to be rational is not the same as assuming it to be true, but is it not the same thing to assume a belief rational as to assume it intelligible? If yes, then the two principles are the same; however, the answer is a flat no.

In trying to maximize rationality in the history of science, it could turn out that making the beliefs of all maximally rational we end up by judging our present beliefs highly irrational. If making the beliefs of others intelligible necessarily means assimilating them to our present outlook, then I do not advocate making them intelligible at all. *R* and the principle of humanity will likely yield similar guidelines in most situations; but, in principle at least, they are rivals.

There is an old view of explanation which had it that to explain is to reduce to the familiar. Such a view is totally untenable in the light of our best explanations of the physical world. Quarks, neutron stars, and DNA are as unfamiliar as anything could be, though they are the stuff of the most wonderful accounts of nature going. The principles of charity and humanity both tell us to deal with foreign beliefs by making them familiar. *R* looks for something common in the beliefs of others and the beliefs of ourselves: it needn't be anything familiar at all. The results of *R* could be as shocking as Freud's psychology or quantum mechanics. The results of employing charity or humanity could never be novel; *a fortiori*, they could never lead us to an *improvement* in *our* beliefs.

Fifth, *R* is supported by the arationality principle. The arationality principle is a methodological principle like the principle of charity. What it does, as I outlined in the first chapter, is to select a preferred explanation from among possible alternative explanations of human actions. The principle says that when given a choice between a rational explanation of some action and some other, such as a psychological or sociological explanation of the same action, the explanation we ought to choose is the rational one. That is, the right explanation is the one where the cause of the action was a good reason. Consider an example. Suppose we have two rival explanations for why Einstein abandoned Newtonian mechanics in 1905. One says that he did so because a rival theory, his own special relativity, better fitted the available evidence. The other explanation says that he abandoned Newtonian mechanics because he was in rebellion against his elders, and so overthrew their most basic beliefs; it was a case of the conflict of generations, an inevitable fight between the son and the father (Feuer 1974). The arationality principle tells us that of these two rival explanations of Einstein's action the one we should choose is the rational one, the one that says he did it for cognitive reasons.

The arationality principle does not say that psychosocial factors never play a role. Often for a given action no rational explanation can be found. Then psychosocial (or other 'external') causes must be appealed to.

In their respective ways these two principles, R and the arationality principle, are trying to do the same thing; they are trying to maximize rationality. They are not equivalent principles; but they are strikingly similar and intimately related. It is because of this relation that I claim support for R from the arationality principle.

The difference between the two principles is this: R is a criterion for choosing among rival methodologies. The arationality principle assumes as already given what rationality is. It is the business of R, for example, to choose between Bayesianism and falsificationism (among others). Only after this choice is made can the arationality principle be employed. Suppose Bayesianism is the winner. Then, given two rival explanations of why a scientist made a certain decision, one saying he (or she) did it because it was the rational (i.e., Bayesian) thing to do, and the other explanation saying that he did it because of his toilet training, then the arationality principle says we should choose the Bayesian explanation. So clearly these two principles work at different levels; but there is a common motivation behind R and the arationality principle, and that is why they are mutually reinforcing. Both try to maximize rationality at their respective levels. Both tend to regard the rational explanation as the best and preferred explanation of any action.

The logical relationship between the two principles seems to be this: what R does is to give the greatest possible scope for the potential application of the arationality principle. By maximizing the coincidence between theoretical and normative reconstructions, R maximizes rationality in the history of science. It chooses that methodology which makes more episodes rational than any other methodology. Moreoever, any theoretical reconstruction which coincides with a normative reconstruction automatically constitutes an explanation of the episode in question. The action is explained as being the rational thing to do; but this, of course, is just what the arationality principle calls for. On the other hand, if the theoretical and normative reconstructions do not coincide for some event, then no choice is possible between a rational and some other type of explanation. No rational explanation exists, according to the methodology in question, and so the arationality principle cannot be invoked in its favour. Thus, when R picks out the methodology which maximizes rationality in the history of science, it inevitably satisfies the aim of the arationality principle. And it does this by giving the principle the largest possible scope for employment.

Some sociologists of knowledge, especially the strong programmers of the Edinburgh school, have complained that greedy rationalists leave only a few crumbs for them. As they see it, the arationality principle, for instance, gives sociologists only the irrational residue to explain. However, this is not the case with R. It is true that sociologists get to explain what is irrational, but that is not their only task. Much more important is the circumscribing (in effect) of the realm of the rational in the first place. R does not just call upon sociologists after rationalists have done all they can; rather, sociologists have a fundamental role to play right from the start. This is quite a different role from that envisaged by either sociologists of knowledge or rationalists in the past. But I'll have a good deal more to say about this below.

I turn now to the disposal of a pseudo-problem. Those who don't see the problem, or who wish to take it on faith that it can be resolved, may wish to bypass the details and skip to the next section; there will be no loss of continuity.

THE PROBLEM OF CIRCULARITY

This alleged problem arises when historical case studies are cited as evidence for a normative philosophy of science. The problematic circularity is seen as much more pressing and important than the normal circularity that arises in any coherentist epistemology. The difficulty goes something like this: Scientists, unlike nature, sometimes make mistakes. There is good science and there is bad science. If a theory of scientific method is to be tested against actual scientific practice (instead of being appraised exclusively by means of a priori considerations) then it must first be settled what are the examples of good and of bad scientific practice. But we cannot classify scientific practice into good and bad except by invoking some theory of scientific method. Thus, testing a methodology by appeal to actual scientific practice turns out to be a circular, question-begging enterprise.

This argument, and variants of it, are often used to show that history of science cannot be used to decide between competing methodologies of science, and that the basis for choosing a normative philosophy of science must be logical, conceptual, and a priori. Even those who *do* wish to use history of science as evidence for or against competing philosophies of science are worried enough by this argument to feel obliged to make proposals which skirt the alleged circularity. The cogency of the circularity argument can be undermined, however, and that is what I propose to do in this section. I will first cite a few different versions of the argument, then the problem will be laid to rest.

Laudan, in discussing a number of difficulties which arise in any attempt to formulate an evidential relationship between the history of science and the normative philosophy of science, expresses the circularity problem this way:

> Foremost among these difficulties is the *vicious circularity* which it seemingly entails. If the writing of history of science presupposes a philosophy of science and if philosophy of science is then to be authenticated by its capacity to lay bare the rationality held to be implicit in the history of science, how can we avoid automatic self-authentication, since the history of science we write will presuppose the very philosophy which the written history will allegedly test?
>
> (Laudan 1977:157)

In spite of this, Laudan is determined to use history of science as evidence for choosing among competing methodologies, and his specific proposals (his PI criterion and his naturalism, both examined above) will, he thinks, avoid the circularity he fears could be present in the situation.

Ron Giere comes to a contrary conclusion. He uses the argument to justify his claim that the history of science can provide the philosophy of science with inspiration and examples only. It cannot provide evidence. It must be on the basis of *a priori* considerations that we choose a methodology. The connection beween the history and the philosophy of science is not 'intimate', writes Giere;[4] it is a 'marriage of convenience'. He puts the circularity argument as follows:

> There is a further problem which is especially acute for any historical approach to criteria for the rational choice of theories.

Suppose . . . that history provides *empirical* data for one's account
of theory choice. In this case the account is itself an empirical
conclusion, or, broadly speaking, a theory. But to choose a
theory of theory choice on the basis of historical data one must
already have some criteria for theory choice.

(Giere 1973:292)

There are other versions of the argument as well (see, for
example, McMullin 1970 or Maull 1976). But the structure of
them all is roughly the same. In order to lay the circularity
problem to rest I will reconstruct the argument in a number of
ways. The presuppositions which are said to hold between the
history of science and the philosophy of science will be understood
sometimes temporally and sometimes logically. Here is one way of
understanding the argument which makes it seem particularly
devastating:

Before (temporally) a philosophy of science can be chosen a
    history of science must be written.
Before (temporally) a history of science can be written a
   philosophy of science must be chosen.

∴ Before (temporally) a philosophy of science can be chosen a
   philosophy of science must be chosen.

The first premiss of this argument says that it is the history of
science which provides the evidence for choosing among competing
methodologies of science and that we must have that historical
evidence before we can make a choice. The second premiss
says – to use the terminology developed above – that the written
history of science is a theoretical reconstruction. Given that
nothing can be chosen before it is chosen, the conclusion would
appear to express an absurdity. Consequently, if the argument is
valid then one or both of the premisses is in trouble.

It would seem reasonable to interpret Giere, Laudan, and others
as understanding the argument in the way just sketched. Their
reaction to it is univocal: they all deny the first premiss. The way
they deny it is somewhat different in each case, however. Giere, for
instance, denies that the history of science can serve as evidence at
all; a methodology must be argued for entirely on a priori grounds.
And on those grounds it can be chosen before the history is written.
On the other hand, Laudan's reaction to the argument (in 1977,

but perhaps no longer) is to break up the history of science into two sets, saying that we have clear 'pre-analytic intuitions' about the members of one of these sets of historical episodes. The competing philosophies of science are to be evaluated on the basis of these, not on the basis of a detailed, theory-laden, or theoretically reconstructed history of science. However, Laudan, Giere, and others have taken the route they have because they wrongly accepted the argument as valid and the conclusion as absurd.

The argument is not cogent; there is an equivocation on 'chosen'. In the first premiss 'chosen' means something like 'accepted' or 'confirmed', while in the second premiss it means something like 'utilized' or 'employed'. If these terms are substituted into the argument in place of 'chosen' we find that the conclusion is a relatively innocuous one:

Before (temporally) a philosophy of science can be confirmed a history of science must be written.
Before (temporally) a history of science can be written a philosophy of science must be utilized.

∴ Before a philosophy of science can be confirmed a philosophy of science must be utilized.

It is not obvious that this conclusion should lead to any further trouble. In fact, it is specifically required by R. In order that we may confirm a normative philosophy of science as better than others, it and its competitors first have to be utilized in the reconstruction of the history of science.

These same premisses seem to be used to draw a different conclusion as well. Laudan worries that the philosophy of science used to do the theoretical reconstruction cannot avoid 'automatic self-authentication'. But his fears are quite unjustified, as can be seen by inspection of the argument when it is laid out.

Before (temporally) a philosophy of science can be confirmed a history of science must be written.
Before (temporally) a history of science can be written a philosophy of science must be utilized.

∴ The philosophy of science utilized will be confirmed.

This conclusion simply does not follow from the premisses. The argument is obviously invalid. The only conclusion which could plausibly be inferred from these premises is the converse of the one drawn in the circularity argument, namely:

The philosophy of science confirmed must be utilized.

This is quite a different conclusion and leads to no trouble. Indeed, once again, it is something specifically maintained in holding R. If a philosophy of science is confirmed it is because it has a greater coincidence of its theoretical and normative reconstructions than any other methodology. Thus, it cannot be confirmed until it has made the historical reconstructions, that is, until it has been utilized.

So far, temporal versions of the circularity argument have been given. Let us look at a logical version now.

Philosophy of science (logically) presupposes history of science.
History of science (logically) presupposes philosophy of science.

∴ Philosophy of science (logically) presupposes philosophy of science.[5]

Let us fill out the premises of this argument with a bit more detail, to see if Laudan is right that we end up with a vicious circle. Suppose H is some account of an episode in the history of science, which is theoretically reconstructed using the conceptual apparatus of some normative philosophy of science P. And further, H is the historical evidence for P.

P is chosen (logically) presupposes H is written.
H is written (logically) presupposes P is chosen.

∴ P is chosen (logically) presupposes P is chosen.

As in the argument above, this is a perfectly harmless conclusion. After all, we are free to infer tautologies with impunity. If we substitute 'confirm' and 'utilize' for the two occurrences of 'chosen' we get, once again, an innocuous conclusion:

P is confirmed (logically) presupposes P is utilized.

We clearly do *not* get the conclusion which would be harmful:

P is utilized (logically) presupposes P is confirmed.

In order to drive this point home I will return to the R account of how it is that history serves as evidence for philosophy. R would seem to imply all of the premises in all of the above arguments. Specifically, that

1. Before (temporally) a philosophy of science can be confirmed a history of science must be written.
2. In order for a philosophy of science to be confirmed a history of science is (logically) presupposed.
3. Before (temporally) a history of science can be written a philosophy of science must be utilized.
4. In order for a history of science to be written a philosophy of science is (logically) presupposed.

Assertions 1 and 2 mean that a methodology is temporally and logically dependent on the historical facts for evidential support. Assertions 3 and 4 mean that all written history of science is theoretically reconstructed.

If the circularity argument is cogent then it should work using these four premises. In order to show that the conclusion

P is utilized (logically) presupposes that P is confirmed.

does *not* follow from these premises it should be sufficient to make obvious that any such conclusion cannot be inferred except with the addition of extra premises saying just how history serves as evidence for methodological theories.

Recall how R says a philosophy of science is confirmed. First, it must be utilized in two ways: to make a theoretical reconstruction (actual history, by its lights) and to make a normative reconstruction (rational history, by its lights). Second, the theoretical and the normative reconstructions are to be compared. The best methodology is the one such that its theoretical and its normative reconstructions of historical episodes more closely coincide than do the reconstructions of other philosophies of science.

Let us add this criterion to the other premises of the argument. Recall that P is some normative philosophy of science and H is a history which has been reconstructed, either normatively or theoretically, using the methodology P. I will distinguish by using '$H(P\text{-theoretical})$' to mean the actual event described through the eyes of the methodology P, and '$H(P\text{-normative})$' for the history as it ought to have gone according to the methodology P, were everything done rationally.

150

1. P is confirmed presupposes $H(P$-theoretical$)$ is written.
2. $H(P$-theoretical$)$ is written presupposes P is utilized.
3. P is confirmed presupposes $H(P$-theoretical$)$ and $H(P$-normative$)$ more closely coincide than do $H(Q$-theoretical$)$ and $H(Q$-normative$)$ for any other methodology Q.

Now, does the problematic conclusion

P is utilized presupposes P is confirmed.

follow from these premisses? Obviously not. For suppose we add another premiss:

4. $H(R$-theoretical$)$ and $H(R$-normative$)$, for some methodology R, more closely coincide than $H(P$-theoretical$)$ and $H(P$-normative$)$.

Then we could conclude that

P is not confirmed.

And this conclusion follows in spite of the fact that P is utilized. This would seem to show that the utilization of a methodology in the reconstruction of history does not automatically lead to its self-authentication.

In sum: I have cited a number of statements of the alleged circularity problem, but the reconstruction of this *prima facie* plausible argument has, in each case, turned out to show that it is not very cogent. In conclusion, then, we can safely say that the circularity problem is a pseudo-problem. It does not stand in the way of using history as evidence for normative methodology, at least not if R specifies how it is that history is to be used for evidence. This is another plus for R.

A WINNER?

Earlier I moved from the inadequacy of sociological approaches to the correctness of a rational view of science. This was a big jump in my argument and I don't want to disguise it. I'm now going to make another big jump, this time to what I take to be the right normative methodology. It is a big jump because my rule R requires an enormous amount of historical and sociological work to justify any normative philosophy of science, and I haven't done that work. Nevertheless, there has been a great deal of work done in the past couple of decades which has been to some extent in the spirit of *R*. For the most part this has been post-Kuhnian work done by rationalist-minded philosophers of science. In broad outline at least, there is considerable consensus. The picture I now paint with broad strokes is not in any way novel; it derives from Kuhn, Lakatos, Laudan, and many others. It, or something like it, I conjecture, is the winner according to my rule *R*.

To start with I'll make a distinction which proves to be very important between theories and something which is more all-embracing, a global unit. This will already be a familiar notion. The idea behind a global unit is that it is something which directs research as well as contains a sequence of individual theories. A global unit (hereafter GU, to be read 'goo', suggesting a sticky substance which holds lots of things together) is akin to Lakatos's (1970) research programmes, Laudan's (1977) research traditions, Holton's (1974) thematas, Toulmin's (1972) intellectual traditions, and, to some extent, Kuhn's (1970) paradigms (understood in the sense of a disciplinary matrix).

A GU is a sequence of different theories. Each theory, after the programme is launched, can be seen as a modification of the one it follows. However, changes within this sequence of theories tend to leave basics unchanged. Individual theories within a GU are detailed accounts of the GU's subject matter. Over time there will be many of them. The ordinary word 'theory' is ambiguous between my use of 'theory' here and a GU. Thus, 'the Copernican theory' could mean either some detailed account given by Copernicus himself and later superseded by the detailed accounts of Kepler and Newton, or it could mean the whole programme of a heliocentric astronomy.

A GU has a set of relatively basic ingredients; these are the core assumptions. For the mechanical philosophers such as Descartes and Boyle the core of their programme was that all phenomena are just the result of matter in motion. Accounts of the nature of light, the motion of the planets, the behaviour of animals, and all other phenomena must be in terms of the core assumption; no appeal to colours and other secondary qualities was permitted. A GU contains not only a sequence of theories but also a number of methodological norms. Though seldom explicitly stated, they nevertheless guide research by stating how the phenomena should be dealt with. This involves telling us how to construct new theories, how to modify old ones, and moreover, how to appraise any existing theory. Thus, the norms of the Newtonian GU tell us to postulate bodies which behave according to his laws of motion and the universal law of gravitation in order to account for planetary motions. The Ptolemaic norms, on the other hand, tell us to try out various combinations of circular motions. And within the mechanical philosophy only contact action is allowed; so, if someone should account for the motions of planets by positing a mechanism of action at a distance, then this would be ruled out as illegitimate, even if it made all the right empirical predictions.

Conceptual issues play a major role. The debates between Newton and Leibniz, and between Einstein and Bohr, for instance, were not over the observable facts. They quarrelled over metaphysical issues. Such debates played a crucial role in the development of physics in both the seventeenth and twentieth centuries. To cite another example, biological theories before Darwin tended to be teleólogical; they postulated biological processes which were goal-directed. The best accounts of physics, however, permitted only so-called efficient causes. So even if teleological accounts of biology were empirically adequate, there remained a recognizable problem with them, a conceptual strain. Darwin's mechanistic account of evolution resolved this conceptual tension and was, accordingly, conceptually superior to its rivals.

The testing process will usually be focused upon some specific detailed theory; but what is really being evaluated is the GU. The aim of each of the various rival GUs is to explain the phenomena and help us to understand how the world works. A GU will put forward a specific theory to do this; but this theory will invariably have several shortcomings. It might only roughly account for the data, it might even contradict them, it might conflict with other acceptable theories, it might violate some of the norms within its own GU, and so on. These shortcomings are anomalies; they are not decisive refutations. There is no such thing as a crucial experiment; that is, there are no tests which deliver a knockout blow to any GU. An anomaly shows the need for a revision of the present detailed version of the GU, but it is no more decisive than that.

Cognizant of these anomalies, the proponents of a GU will put forward another version which tries to overcome the deficiencies of the previous theory. This is the key to the whole appraisal process, since modifications can be either progressive or *ad hoc*. Should novel predictions (predictions of a phenomenon not already known to obtain) pan out experimentally, or should a significant conceptual simplification be introduced, then we would look quite favourably upon the GU as a result of this progress. On the other hand, a GU which is just able to keep up with innovations and discoveries coming from its rivals is one we quite rightly think of as degenerating, and typically we label the moves made by its champions as *ad hoc*.

There are several different cognitive stances one might take toward any GU. An obvious one is to simply accept or believe it. This is just to accept or believe the latest detailed theory in the GU. But it is important to note, as stressed by Laudan (1976), that there are other cognitive positions as well. The most important of these is the stance of pursuit-worthiness. Often we assess a GU as 'promising', even though it is not acceptable as it stands. We are prepared to spend some of our finite resources on developing it even though at present we believe some rival theory more likely to be true. Once we make this distinction between acceptance and promise, it follows that it might have been rational to accept Ptolemy but pursue Copernicus in the sixteenth century. By the eighteenth, however, it had become rational to both pursue and accept Copernicanism.

A crucial consequence of the foregoing sketch of scientific rationality is this: theory choice is basically comparative. To be rational is to choose the best of the lot. There is no question of an absolute standard of success or failure that any theory can be compared with. A theory can only be compared with its rivals. Positivistic notions of confirmation and Popper's falsificationism are both absolutistic accounts of rationality. On both of these we can take a single theory to nature for judgement, and nature will say 'yes' or 'no'.

On a comparative account, nature will say nothing about a single theory. What nature does is rank-order a set of theories. A swimmer or a runner can be tested, I suppose, against the clock; but there is no way to evaluate the merits of solitary boxers. They can only be tested by tossing them into the ring with other boxers. Evaluating a particular boxer by rank-ordering the set of boxers can be a perfectly objective process, but it is fundamentally comparative. Evaluating a scientific theory by rank-ordering the set of rivals is similarly comparative; nevertheless, it is objective. The fact that theory choice is fundamentally comparative has significant consequences for the sociology of knowledge. This has recently been stressed and clarified by Kathleen Okruhlik, whose work I now follow.

DISCOVERY VERSUS JUSTIFICATION

The lore of science contains many an interesting anecdote about how theories came into being: an apple fell on Newton's head, a snake bit its own tail in Kekulé's dream; and thus were born the theories of universal gravitation and the structure of the benzene molecule. Of course, thinking up a theory in the first place is one thing; having a good reason to believe it is quite another. There is danger, we might suppose, in running them together. 'The only way to escape this difficulty', says Reichenbach, 'is to distinguish carefully the task of epistemology from that of psychology . . . I shall introduce the terms *context of discovery* and *context of justification* to mark this distinction. Then we have to say that epistemology is only concerned [with] the context of justification' (Reichenbach 1938:5–7).

Reichenbach had quite detailed views in the logical positivist mould of just how any justification would go. Popper was adamantly opposed to such an inductivist outlook; nevertheless, he was of one mind with Reichenbach about the distinction, merely substituting his falsificationism when it came to providing the details.

> The initial stage, the act of conceiving or inventing a theory, seems to me neither to call for logical analysis nor to be susceptible of it. The question how it happens that a new idea occurs to a man – whether it is a musical theme, a dramatic conflict, or a scientific theory – may be of great interest to empirical psychology; but it is irrelevant to the logical analysis of scientific knowledge. There is no such thing as a logical method of having new ideas, or a logical reconstruction of this process.
>
> (Popper 1959:31f.)

It's a small irony that Popper's *Logic of Scientific Discovery* so emphatically denies the subject matter of its title.

For two generations the distinction has achieved near-universal assent (Hanson (1961) being a notable dissenter). The reason for its wide appeal is obvious. Even though it gave up a major part of the process of science to non-rational factors, these factors got filtered out. Popper and the positivists disagreed on the filters, but they certainly concurred that filters were there and that they worked. It did not matter what kind of bizarre ingredients went into theory discovery, since they all came out in the wash of justification. It didn't matter that X's theory was born of racial hatred and Y's of a gushy universal love; both could be taken to the tribunal of nature, which would provide a dispassionate 'yes' or 'no' in either case. Thus, social factors could not make it into the *content* of any properly tested theory.

Now all of this was well and good in the days when theory choice was thought to be absolute. But things have changed. As I outlined above, theory choice is comparative, with the consequence that justification, even when it is working perfectly, is not the perfect filter. The reason is very simple. Any social or psychological factor which is systematically present in the comparison group will be systematically overlooked in the process of rank-ordering.

> If our choice among rivals is irreducibly comparative . . . then scientific methodology cannot guarantee (even on the most optimistic scenario) that the preferred theory is true – only that it is epistemically superior to the other actually available contenders. But if all these contenders have been affected by sociological factors, nothing in the appraisal machinery will completely 'purify' the successful theory. . . .
>
> Every choice among alternatives can be a rational choice. Science can (in principle) get better and better. But this in no way guarantees that the content of science is insulated against social influences. Once you grant that social factors may influence the context of theory generation, then you have to admit that they may also influence the content of science.
>
> (Okruhlik forthcoming, a)

Okruhlik gives a couple of examples to illustrate this. She asks us to suppose that the rival hypotheses in cancer research are environmental and genetic. Now if the cancer research were to be funded largely by manufacturers of alleged carcinogens, there would be a strong influence toward the generation of theories which posit non-environmental factors as the cause of cancer. 'If so, the machinery of appraisal might well favour these hypotheses over their underdeveloped rivals; and, of course, hypotheses that don't even get formulated never enter into the picture at all' (ibid.).

A second example concerns the history of theories of female behaviour.

> These theories may in many respects be quite different from one another; but if they have all been generated by males operating in a deeply sexist culture, all will be contaminated by sexism. Non-sexist rivals will never be generated. Hence, the theory which is selected by the canons of scientific appraisal will simply be the best of the sexist rivals. And the very *content* of science will be sexist – no matter how rigorously we apply objective standards of assessment in the context of justification.
>
> (Okruhlik forthcoming, a)

Okruhlik's argument, as she notes herself, is remarkably simple. However, its consequences are striking. It was just a matter of putting the discovery/justification distinction together with comparative models of theory evaluation. The effect is that we must now completely rethink the role of the social in the very content of science, for the rules of rationality, even when flawlessly employed, may sometimes be incapable of filtering out the social.

Ian Hacking has come to similar conclusions. In his (1986) essay on weapons research he argues that the massive amount of military funding has actually shaped the very content of all scientific knowledge. There is an old rationalist view which went as follows: There may have been a military influence on the direction of Galileo's research so that he looked into the motion of cannon balls rather than into some other phenomenon; but there is no external influence on the content of his findings, namely that the trajectory of a cannon ball is a parabola. This, it seems, is somewhat naïve. What Hacking and Okruhlik argue is that direction of research can influence content, even when only rational choices are made among rival theories.

Ferreting out systematic social influences in the comparison group is one more major task for the sociologist of science. The extent to which social factors play a role in generating the comparison group is an almost complete unknown. How should sociologists tackle the problem? For one thing, they can start looking for correlations. As I mentioned above, Shapin has already stressed this feature, and I completely agree with him. Beyond this homely advice it is up to sociologists themselves to find their own way. But this I do readily concede: their findings are sociological findings, yet they are central to our understanding of how science works. It is part of the complete anthropology of science; it is part of the total rationalist image.

CONCLUSION

So what now is the job of the anthropologist of science? It is, of course, to understand how science works. The kind of anthropological approach to science which I advocate will do a lot of explaining in terms of good reasons, evidence, and so on. But there will also be a big sociological component. The sociological aspects include the following:

1. To determine general sociological truths. (For example, the Matthew Effect, or that judgements are influenced by peer groups, or that people desire fame and fortune, or that members of some races, sexes, or classes would like to lord it over others; and so scientists will often act in deviant ways to achieve their goals.) These general sociological truths act as a constraint on the degree of rationality we should attribute to the entire history of science.

2. Once we have a theory of rationality, we use it to explain events in the history of science. Irrational beliefs are to be explained by appeal to social (or psychological or other) forces. (Bloor complained that rationalists give sociologists mere crumbs, only the irrational residue to explain. On my view these crumbs remain, but I'm adding a great deal more. It is not at all what he was expecting, but Bloor should nevertheless have a very substantial feast.)

3. Rational evaluation is fundamentally comparative; consequently, social factors may be systematically present in the comparison group and so make their way into the very content of the best theories. It is a sociological task to determine the existence of such systematic social factors.

The anthropology of science I advocate takes all three into account. And once we understand these features of science, we can begin to think about improving it.

Chapter Seven

MAKING SCIENCE BETTER

There is something intriguing about strong-programme claims to be explaining science, not evaluating it; its advocates see themselves as admiring science, not sabotaging it. Sociological explanation, says Bloor, is just the scientific (hence proper) way of understanding science. There is usually not a hint of criticism about any scientific venture. In the same spirit, Collins laments the fact that through a lack of understanding people often turn away from science, and indeed, turn against it. ('A loss of confidence in the scientific enterprise is a disaster that we cannot afford' (1985:165).) On the other hand there has been a large number of feminist discussions of science recently, and they tend to be highly critical. If some of these critics are correct, significant chunks of science are exposed as a fraudulent, gender-biased activity. In so far as these attacks are successful, the authority of science is seriously undermined in consequence.

Strong programmers and feminists alike give sociological explanations for various scientific activities; but one group sees this as supportive of ordinary science while the other sees it as undermining. Who's right? Well, of course, the feminists are. Once a belief is seen to be due to social forces its credibility is diminished; once it is established as ideology its plausibility takes a tumble. If Forman had convincingly established his account of the rise of non-deterministic quantum mechanics, and had done so during the very period of its rise, who would then have continued to believe in the new quantum theory? If people now believe in the truth of quantum mechanics and also in Forman's account of its rise, then it is because they think they have independent evidence for the truth of the theory. I need not speak as a rationalist here;

whether people are rational to act this way or not, it is nevertheless a psychological law that *when convinced that a belief has been held because of sociological factors, people tend to lessen its credibility.* Whether they want to or not, sociologists undermine the authority of science.[1]

The crucial difference between recent feminist analyses of science and the strong programme is whether science itself is open to criticism. Showing that social factors are responsible for a belief is merely the first step for a feminist critic: ideally it leads to the rejection of that belief and its replacement with a better one. This is exactly the right attitude. Ideology, once exposed, is not to be gazed upon like other flora and fauna; it is to be stamped out. Paraphrasing Marx, strong programmers have only interpreted science; the point, however, is to improve it. This basically philosophical task is a kind of normative intervention in the scientific enterprise. Philosophers, of course, are incurable moralizers, and science is an opportunity not to be missed.

BARNES ON IDEOLOGY

Barry Barnes wants to save and utilize the concept of ideology. At first blush this seems strange; after all, is not ideology just a belief which arises as a result of interests rather than objective evidential considerations? Barnes, a paradigmatic strong-programme sociologist, claims that all beliefs arise from interests; so how can he maintain a distinction between 'good' science and 'ideology'?

According to Barnes, a distinction can be made between different sorts of interests; this distinction is the basis for separating ideological from non-ideological beliefs. The division is between concealed interests, on the one hand, and our general interest in prediction and control, on the other. He writes:

> whenever knowledge is ideologically determined there is disguise or concealment of an interest which generates or sustains the knowledge, or, to put it another way, of the problem to which the knowledge is actually a solution. This gives us a basis for the definition of ideological determination. Knowledge or culture is ideologically determined in so far as it is created, accepted or sustained by concealed, unacknowledged, illegitimate interests.
>
> (Barnes 1977:33)

Given this characterization of ideology,[2] a rather obvious question arises: How do we test for ideology? How can we identify a piece of pseudo-science? Barnes answers this question by proposing the following method:

> it remains possible in many instances to identify the operation of concealed interests by a subjective, experimental approach. Where an actor gives a legitimating account of his adhesion to a belief or a set of beliefs we can test that account in the laboratory of our own consciousness. Adopting the cultural orientation of the actor, programming ourselves with his programmes, we can assess what plausibility the beliefs possess for us. In so far as our cognitive proclivities can be taken as the same as those of the actor, our assessment is evidence of the authenticity of his account.
>
> (Barnes 1977:35)

Let us look at each of these proposals in turn: first the definition of ideology, then the criterion of applicability. Barnes seems to be unique among cognitive relativists for wanting to do some sort of justice to the concept of ideology and not just pass it off as a philosopher's illusion. Rationalists cannot but applaud his good sense on this score. Indeed, as we shall see momentarily, rationalists can even concur with his concrete suggestions, but I doubt whether Barnes himself can. His account of ideology is totally at odds with everything else he subscribes to about the nature of scientific belief; in other words, it is incompatible with the strong programme.

Given his relativism, one wonders why Barnes has used the terms 'legitimate' and 'illegitimate' in describing the two different kinds of interests which might produce belief. These are normative words, the kind of terms sociologists of knowledge usually shun in their search for 'naturalistic' accounts. However, the terms do not play any great role in the definition of ideology and it is not a question we need worry about here.

A truly serious problem for Barnes's account of ideology is this: it makes all the case studies which are used to support the strong programme turn out to be instances of ideology. Consider some of the examples. Forman's study of causality and quantum mechanics claims that the Weimar scientists were out to regain the prestige they lost as a result of the defeat of Germany in the First World

War. The German public was taken with chance; so the physicists, according to Forman, got rid of determinism in the quantum theory. Since the interests served by their actions had to do with their own prestige and not with a general interest in prediction and control, it follows by Barnes's definition that this is a case of ideological determination.

In his study of the Edinburgh phrenology debates, Shapin appeals to middle- and upper-class interests, respectively, to explain the positions of the two warring parties. Once again, beliefs are determined by narrow interests (middle- and upper-class), not a general interest in prediction and control. Finally, Pasteur's reaction to the theories of Pouchet on spontaneous generation are attributed to Pasteur's extreme right-wing political views by Farley and Geison in their study of the debate. Again, Pasteur's narrow and concealed interest in preserving the political status quo is responsible for the belief.

In each and every one of the case studies so often cited by cognitive sociologists as supporting the strong programme, the beliefs in question turn out to be ideological beliefs by the characterization that Barnes gives of ideology. We can only wonder what a sociological account of non-ideological beliefs might look like.

Of course, this situation, which is a quite surprising consequence of Barnes's characterization, is nevertheless perfectly congenial to a rationalist. The rationalist has all along maintained that ideology should be given a sociological explanation. This is the essence of the arationality principle. Rationalists will characterize bad science, or ideology, as the product of not following the proper rules of scientific conduct. As a factual claim about why there is deviation from the proper rules, Barnes's characterization is certainly an acceptable conjecture, and even quite plausible. I will turn now to Barnes's criterion for identifying ideological belief. His proposal is that we should adopt the cultural orientation of the actor, programme outselves accordingly, and see whether the results have any plausibility for us. The resulting belief is not ideology if and only if it is plausible to us.

That such a proposal should come from Barnes is quite remarkable. Within Barnes's general framework it is a totally unworkable procedure (yet there is really nothing else he could say). The first consideration to bring to bear is the obvious fact that the conclusion depends on the testing agent. If the agent who is doing the subjective thought-experiment is also hiding or failing to acknowledge the same special 'illegitimate' interests, then that agent will come to the same conclusion as the original actor whose beliefs are being tested for ideological determination. On the other hand, if the investigator has special interests that the actor under investigation does not have, then the result of the experiment will be the incorrect conclusion that ideology is at work because there is a divergence in conclusions.

There are two responses that Barnes might make to this. One is that in the testing procedure only bona fide judgements are to be allowed. But often people are the unwitting holders of ideological beliefs; people do not always consciously conceal special interests, but they may have them, nevertheless. The flawed judgements that they make can be made in perfectly good faith. A second response that might tempt Barnes is to try the democratic route: a belief is ideological if it diverges from the conclusions of the majority. Unfortunately, counter-examples abound. By any standard of ideology, eighteenth-century opinion on slavery (or rather, on the biology and psychology of Africans) was ideology, even though it may have been majority opinion (among whites).

Though Barnes's proposal for identifying ideology is completely unworkable within his general framework, that is, it is unworkable within a thoroughgoing sociology of knowledge, still it is not such a bad procedure. When faced with the kind of difficulties which I have pointed out within his proposal, the temptation in each case is to say something like this: The testing agent who starts from the initial point and proceeds correctly is the right sort of agent for discovering whether a consequent belief is or is not ideology. This, of course, is a temptation that any rationalist could cheerfully give in to, but it is not one that Barnes could accept. The reason that he could not accept it is that to proceed correctly is to proceed rationally, or in accord with the rules of proper method. This he thinks is impossible because he denies the existence of such correct methodological rules.

Given the obvious failure of Barnes's sociological account of ideology and given the equally obvious ease with which any rationalist view of science can make sense of the notion, here is a perfectly plausible conclusion. The successful criticism of any piece of science requires a rationalist outlook. It is a common caricature of rationalists that they think everything done in the name of science is rational. Not only is this dead wrong, but, moreover, it is *only* the rationalist who can recognize the very existence of bad science. Consequently, it is only the rationalist who can make it better.

HARDING'S FEMINIST APPROACHES

Sandra Harding's *The Science Question in Feminism* (1986) is typical of much recent feminist criticism of science. On the one hand, it is highly critical. Harding rightly notes that science has contributed to the suppression of women; she cites a number of examples, then remarks that 'all these have been justified on the basis of sexist research and maintained through technologies, developed out of this research' (1986:21). But on the other hand, she is not dismissive of science, though the reforms she calls for are radical. As Harding herself puts it, she is 'seeking an end to androcentrism, not to systematic inquiry' (1986:10). (See Okruhlik forthcoming, a and b, and Wylie 1987.)

Harding outlines three quite distinct approaches to science that feminists might try. Each of these feminist epistemologies constitutes an analysis of what is wrong with science and how it can be improved.

Feminist empiricism holds that there is in existence such a thing as good scientific method. Sexist science is the result of not following the canons of rationality, and it could be eliminated if proper scientific method were only followed more scrupulously. The social situation of the scientist is not epistemologically relevant.

Feminist-standpoint epistemology holds that the ruling ideas are the ideas of the ruling class, as Marx once put it. But the beliefs of those in power are corrupted by that power. Suppressed groups such as women, the working class, and racial minorities have a clearer vision of how things really are. Thus, since they are relatively unbiased by power, women (in principle) can do better science.

Feminist postmodernism is a form of sceptical relativism. Theories are understood not to be objectively true or false, rational or irrational.

They are tools for promoting social and political goals.

Harding sees all three approaches as non-sexist, but, reading between the lines, I'd say she favours the last. This, I think, is a mistake. The position argued for in the preceding chapter is some sort of combination of feminist empiricism and feminist-standpoint epistemology. Let me try briefly to defend this now.

There are two complaints Harding has with feminist empiricism. One sort of complaint has to do with the well-known problems of any sort of empiricism. But such a criticism seems quite unfair. 'Feminist empiricism' is just her name for what is taken to be the existing canons of rationality; they needn't be empiricist. Popper might claim that proper science is done by the method of conjectures and refutations and that sexist science is the result of trying to 'confirm' rather than 'refute' our theories. As such he would be a feminist empiricist, though not an empiricist.

Harding's other objection is this: why should we believe there is such a thing as actual canons of good science if they are so easy to ignore? This is a much harder challenge to meet. What I will venture to say is that the rules of good science are not flouted anywhere near as often as she thinks. But since there are so many examples of sexist science, doesn't this claim obviously fly in the face of the facts?

Return now to the results of the last chapter. There it was argued that we should interpret the history of science to make it as rational as possible, subject to independent sociological findings. But rationality, remember, is comparative; it consists in making choices from the available options. If all the rival theories are generated by sexists, then even flawless rationality will not be able to filter the sexist content out of the result. The best theory will still be sexist. With this in mind, we can now answer Harding's question. Sexist science results, not only from ignoring the canons of rationality, but from having a systematically contaminated comparison group to choose from. Of course, some will flout the rules of good science deliberately to achieve a sexist outcome, and I don't doubt that this is sometimes effective. However, the real problem is with the comparison group, and this brings me to Harding's own criticisms of the view she calls the 'feminist standpoint'.

Suppressed groups, according to this view, see more clearly and are thus capable of doing better science. The problem with such a

165

view is that there are several quite distinct suppressed groups. As Harding points out, white middle-class women are suppressed as far as their sex is concerned but are part of the suppressor group when it comes to class or race. So what about the standpoints of blacks, the working class, and so on? It seems hard to maintain, says Harding, that there is one correct standpoint that truly sees things clearly.

In answer to Harding's criticism of the feminist standpoint, we should concede that she is right; there is no single correct standpoint. But that is not important. Any standpoint epistemology is good simply because it is different. All standpoints should be encouraged simply because they enrich the comparison group.

The view which I have tried to outline (based in part on Okruhlik forthcoming, a and b) is a blend of Harding's first two feminist epistemologies. It is akin to feminist empiricism in holding that there are existing methods of good science and that they should be maintained and rigorously applied. It is also akin to the feminist-standpoint view, not in thinking a priori that women's perceptions are better, but in thinking that they are likely to be different, and so that the theories they generate might prove to be better when compared with all their rivals.

On such a view science can make progress, which it cannot on the sceptical relativism of nihilistic postmodernism.

EXAMPLES

In the last chapter I outlined different ways social factors can get into science; discovering these is part of the job of the 'anthropologist of science'. In this chapter I am interested in how to make science better. The two are connected. Discovering social factors that possibly contaminate science is the first step in eliminating them. Let me briefly describe some instances of both, which are drawn from recent feminist literature.

Evelyn Fox Keller in her interesting biography of Barbara McClintock, *A Feeling for the Organism* (1983), describes a classically trained scientist who tends to go her own way, a way which is quite different from that of her male colleagues. Keller suggests that McClintock's 'feeling for the organism' was typically female in that it was much more 'holistic'. Because of this fact she was able to imagine processes of inheritance in the corn she was studying that her more reductionistically minded male colleagues could not conceive.

Of course, this is somewhat speculative; but if Keller is right, then this is a fine example of the essentially sociological task of determining the sources of the rival theories in any comparison group. It also has clear normative force in that obviously genetics was improved by having a different kind of person (i.e., a woman) contribute to the pool of alternatives.

Ruth Hubbard has drawn attention to an example that would be amusing if it were not for its pernicious consequences. The ethologist Wolfgang Wickler, writing about sexual behaviour, remarks:

> Even among very simple organisms such as algae, which have threadlike rows of cells one behind the other, one can observe that during copulation the cells of one thread act as males with regard to the cells of a second, but as females with regard to the cells of a third. The mark of male behavior is that the cell activity crawls or swims over the other; the female cell remains passive.
>
> (Cited in Hubbard 1982:31)

Hubbard goes on to give a brief analysis which is exactly right:

> The circle is simple to construct: one starts with the Victorian stereotype of the active male and the passive female, then looks at animals, algae, bacteria, people, and calls all passive behavior feminine, active or goal-oriented behavior masculine. And it works! The Victorian stereotype is biologically determined: even algae behave that way.
>
> (Hubbard 1982:31)

The accounts of Keller and Hubbard may be arguable, but others are not. The French anthropologist Claude Lévi-Strauss once described a village as 'deserted' when all the adult males had vanished. No woman in the same circumstances would have said: 'The entire village left the next day in about 30 canoes, leaving us alone with the women and children in the abandoned houses' (cited in Eichler and Lapoint 1985:11).

These examples just skim the surface of the increasingly rich feminist literature on science.[3] By endorsing them I am not suggesting that women are better scientists than men, or that women have no biases. Rather, they have different ones; and many of these biases (male or female) will become apparent only by contrast. The more we have to contrast any theory with, the better we are able to evaluate it.

I want to pursue the question of intervening to make science better, not by citing additional examples, but in a more general setting. For the balance of this chapter I will focus on the special problem of observation and experimentation, since it undoubtedly plays such a big role in scientific practice. The stage will be set with a few remarks on 'naturalism'.

NATURALISM

Increasingly, approaches to knowledge today are naturalistic: or at least a growing number of accounts of perception, cognition, observation, inference, experience, and other aspects of epistemology are called 'naturalistic' by their advocates, (though often this common label leaves more diversity concealed than similarity revealed). Among those who have adopted the appellation are such rationalists as Ron Giere (1984), Larry Laudan (1984, 1986, 1987), and Dudley Shapere (1982); also included are those who take a very strong lead from Darwin: Willard Quine (1969a, 1969b), Donald Campbell (1973), and Michael Ruse (1986). Perhaps the most outspoken naturalists are the strong-programme sociologists whose views we have been examining throughout this book.

It is easy enough to see the appeal of naturalism: because of its vagueness, the term readily applies to just about everyone in some sense or other, and who, after all, wants to be *un*-natural? Like fibre in the highly touted natural diet, the term 'naturalism' may contribute nothing in the way of nutrition; but it makes the view easier to digest. A term with such wide application borders on the vacuous, however; so let's try to make it precise.

One way to characterize epistemology naturalistically is to say that the one and only way to know is the scientific way. This is how one advocate recently put it (Stroud 1981). But this formulation won't cut much ice, since just about all of us can agree with it. A thoroughgoing Cartesian – at least as popularly characterized – is an apriorist in philosophy just as in physics. And David Bloor, as we have seen, characterizes his strong programme as being just the scientific approach to understanding science; yet his rationalist rivals, say, Lakatos, Laudan, or Newton-Smith, as often as not, see one and only one method of being rational, whether it be in science, in literary criticism, or in philosophy itself.

Let's try another approach. We usually think of the natural realm as being just the physical universe which consists of material objects in space and time. Events and processes consist of these objects causally interacting with one another. To explain any natural phenomenon is to cite the natural processes and events which caused it. The phenomenon that we are interested in here is the development of knowledge, so a naturalistic account of it must be in terms of natural processes. The crux of this form of naturalism is what it means to exclude: norms and other abstract objects are denied any significance in the development of knowledge. They are either denied existence outright, or they are said to be non-natural human artefacts, or they are declared to be just so many epiphenomena which float above and do not take part in the real causal processes of nature. Whatever norms are, they play no role in the natural process of belief-acquisition.

Other things which naturalists typically reject are any sort of apriorism or conventionalism. Though these are important aspects, it is, nevertheless, the denial of a role for norms that I want to focus on here. Any sort of normative evaluation in epistemology is what strong programmers especially reject. Barry Barnes, for instance, takes naturalism to be the employment of the principle of equivalence, which, as we saw above, is essentially the same as Bloor's symmetry principle. He concerns himself with 'the naturalistic understanding of what people take to be knowledge, and not with the evaluative assessment of what deserves so to be taken' (1977:1).

I am going to take issue with this epistemological naturalism, in either its rationalist or its sociological form, especially as this naturalism pertains to experimental observation and sensory experience.

SHAPERE ON OBSERVATION

Dudley Shapere has recently given an interesting and quite sophisticated account of what an experimental observation is; in particular, of what a *direct observation* is:

x is directly observed (observable) if:

1. information is received (can be received) by an appropriate receptor; and
2. that information is (can be) transmitted directly, i.e., without interference, to the receptor from the entity *x* (which is the source of the information).

(Shapere 1982:492)

The account is nicely illustrated with a detailed examination of the solar neutrino experiments wherein the claim is made, both by Shapere and by the participating scientists themselves, that the interior of the sun is being 'directly observed'. This is not as counter-intuitive as it first seems, since Shapere makes a sharp distinction between observation and sense-experience. Sometimes observation and sense-experience will coincide, as when a human is the receptor referred to in the above definition; but cameras, tanks of cleaning fluid, and other recording devices can observe in Shapere's sense without having any sense-experience at all. For this reason observation is a well-understood physical process (at least it will be as well understood as the underlying physical process is); and so Shapere's account is quite appropriately thought of as a naturalistic account of observation.

Shapere likens his own view to that of Quine. Among other things, Quine holds (see Quine 1969a, 1969b) that we have an innate tendency to classify according to qualities such as colour. The hallmark of mature science is in the move away from this to logical and mathematical representations of the world. This, of course, is reminiscence of the view that an earlier age called the primary/secondary property distinction. We also find it to some extent in Shapere's claim that as science develops so grows the distinction between experience and observation, so that now there is often a considerable gulf between them.

There is much to applaud in Shapere's account, but there are also some shortcomings. To start with, the notion of 'information' might prove problematic in Shapere's definition. Information is to pass from the source to the receptor, but does 'information' just mean a causal signal? If so, then it would seem that grass can observe the sun, that rocks can observe the earth, and that my shoes are at present observing my feet. In each case, the grass,

170

rocks, and shoes, receive causal signals from the sun, the earth, and my feet, respectively.

This is no problem, it would seem, if we are prepared to make a distinction between an observation and what is *known* to be an observation. Shapere says that to know (or at least believe) that an observation has taken place we must have theories of the source, the transmission, and the receptor. In the case of the solar neutrino experiments we have theories about how neutrinos are emitted from the interior of the sun, how they get here from there, and how they interact with tanks of cleaning fluid far below the surface of the earth. We are justified in holding that an observation has taken place because we are justified in holding these theories.

However, if this account is right, then we may be led to a bizarre consequence. Knowing what we now know about Jupiter and its moons, the transmission of light, the nature of the telescope, and the physiology of Galileo's eye, we can say that Galileo *observed* the moons of Jupiter. But, since Galileo himself lacked any sort of appropriate optical theory, we cannot say that he realized that he was observing them. To know that one is observing, on Shapere's account, one must know what the underlying physical process is. Galileo didn't. However, Galileo did recognize that he was *experiencing* the moons of Jupiter.

Shapere remarks that on his account, observations are theory-laden. But I doubt that this is true. It is *our conception* of which processes are observations that is theory-laden, not the processes themselves, which occur quite independently of our beliefs about them. Cameras take pictures and thus observe without knowing a shred of optics. If anything is theory-laden it must be experience, not Shapere's observations.

Blondlot, we are now quite certain, didn't observe N-rays. (For a brief account see Klotz 1980.) We are sure he didn't because we're sure there aren't any. Nevertheless, he seems to have had some sort of experience, some sort of appropriate sense-perception. (Either that or he and several of his colleagues were just outright frauds, which we can safely doubt.) Perhaps sensory experience is playing a bigger role than Shapere thinks.

One of the things that Shapere fails to take notice of is the *relativized* notion of observation that is often at work in science. One can think of scientific inquiry as proceeding at various levels. At the macro-level, tables and chairs are observed while electrons are inferred entities; at the high-energy level, electrons and protons are observed while quarks are treated as theoretical. True, it is often very useful to talk in a fashion which suggests that electrons or the interior of the sun are observed. But such observations are really just relative observations; and such talk is parasitic on a form of observation which is not relativized.

Finally, Shapere wants a sharp distinction between observation and sense-experience, and I don't want to deny it. But he also wants the whole burden of experimental evidence, which plays such an important role in science, to fall entirely on observation. Sense-experience is given the brush-off. The problems I have pointed out with Shapere's account suggest, on the contrary, that there is still much to concern ourselves with in sense-experience. This, of course, is where I get my foot in the door. Shapere's receptors just receive the causal signals transmitted from the source. The whole process of observation is a natural one, completely value-free. Cameras and tanks of cleaning fluid, after all, don't have philosophical beliefs. But when we introduce human sense-perceivers we may well have to tell a different story, one in which non-natural factors play a major role.

FODOR AND HACKING ON THEORY-LADENNESS

In an article in which he unabashedly desires to put back the clock, Jerry Fodor asserts that 'there is a theory neutral observation/inference distinction' (1984). This really amounts to two distinct claims. One of the claims is that there is a distinction between what is observed and what is inferred or theoretical. (Here I am following both Fodor and established tradition in using 'observation' and 'experience' to mean the same.) In this regard he is quite convincing, but then hardly anyone believes otherwise. Some, such as Grover Maxwell (1962), have claimed that everything is in principle observable. However, most philosophers nowadays think that the boundary between what is observed and what is theoretical is fuzzy, but, nevertheless, that there are poles with the definitely theoretical and the definitely observational at the extremities.[4]

Fodor's other assertion is that the distinction is theory-neutral. But that claim is not argued for at all; in fact, Fodor even seems to concede that the claim can't be justified. What he gives us in its place is the 'modular' mind, a compartmentalization of cognition. This lends support to a *relative* neutrality in experience: 'if perceptual processes are modular, then, by definition, bodies of theory that are inaccessible to the modules *do not affect the way the perceiver sees the world*' (Fodor 1984:38). Ian Hacking makes a similar claim, playing on the 'disunity of science' rather than the 'modular mind'. 'It is precisely the disunity of science that allows us to observe (deploying one massive batch of theoretical assumptions) another aspect of nature (about which we have an unconnected bunch of ideas)' (1983:183). The point for both Fodor and Hacking is that we have *relatively neutral* observations; the observations under consideration may be theory-laden, but not with the theory being evaluated.

There is a powerful case to be made, I think, that the sorts of relatively neutral observations envisaged by Fodor and by Hacking are at work much or even most of the time in science. The extreme claims of a thoroughly conditioned experience made by Hanson (1958), Feyerabend (1975), and Kuhn (1970) are overstated. But there are cases, on the other hand, where we don't have neutral observations at all, not even relatively neutral ones. Let me illustrate with an example borrowed from Churchland (1979).[5]

Imagine a society of people just like us in their physiology and in their language except that where we speak of heat, they talk of caloric. All material bodies, according to them, hold caloric. The details include the following. The amount of caloric in any given body can vary. When the caloric pressures differ it will flow from one body with the higher pressure to another of lower pressure with which it is in contact; it is distributed evenly throughout any given body. One can feel caloric at various pressures, but when the pressure is great enough one can see it as well, since it is a glowing red. (At even greater pressures it becomes white.) Where we would say a body is hot or cold, they will say it is at high or low caloric pressure.

We, of course, take heat to be observable stuff, and we try to explain heat phenomena by introducing such explanations as caloric or the kinetic theory. Our imaginary society takes caloric to be as observable and as much a matter of common sense as anything could be. Quite possibly they are interested in explaining its properties and ascertaining its microscopic nature, but they have no doubt about its existence; after all, they can see it!

TWO TYPES OF EXPERIENCE

I want to introduce a taxonomy. On the basis of the foregoing two sections there seem to be two distinct types of observation or experience: those that are relatively neutral and those that are not.

> *Type I*: These are observations/experiences that are theory-laden, but they can be tested and either supported or overruled by neutral observations; that is, they can be evaluated by *other* observations, which are neutral with respect to the experiences in question.

Let us consider a couple of examples. The famous Müller-Lyer arrows provide an interesting illustration.

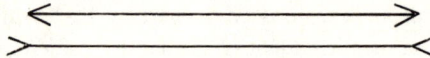

Our perceptual experience is of lines of different length; the top one looks shorter than the bottom one. There are standard psychological accounts of why we have this illusion, which have to do with our (subconscious) beliefs about inside and outside corners, and so on (see Gregory 1970 for a brief account). The important thing, however, is that we have a way to evaluate this experience; that is, we have a way to tell whether or not we should believe what we see. The method is just to hold a ruler alongside the two lines and read off their respective lengths. Of course, this in itself may be a theory-laden observation, but it is one which is relatively neutral.

Often the theory-ladenness of our perceptions is the result of social forces and cultural values. Apparently, poor children perceive the relative size of coins as being larger than their middle-class counterparts do. And when women interrupt a conversation it registers more strongly than if a man interrupts. Typically, men interrupt about 90 per cent of the time in their conversations with women, yet should a woman interrupt more than her standard 10 per cent she is seen by most outside observers as being verbally aggressive and as being the one who did most of the talking (see Ayim 1982 for discussion of speech patterns).

These are both type-I experiences, since we have relatively neutral methods of ascertaining the veracity of those perceptions. In the one case we can simply hold the coins and the other objects they were compared with together, to see which is the larger. Given this opportunity even the poor children who made the initial size ranking based on their experience of the objects flashed on a screen will overrule their initial perceptual beliefs. In the second case, we can give outside observers of the conversation a counter to keep track of interruptions and let them listen to the conversation again (supposing we have it taped). The initial impression of who interrupted whom more often will be changed. Of course there are all kinds of assumptions involved in carrying out these measurements, but the observations used for testing are, with respect to what is at issue, neutral. They are relatively neutral in the way that both Fodor and Hacking want. The moral for experimentation is obvious. Any experimental observation which is type I is a relatively neutral experimental test.[6] But what about the second sort of experimental observation?

> *Type II*: These are observations/experiences for which there is no relatively neutral observation (at least at the time the observation is made), which can be employed to evaluate the veracity of the perception.

Two questions immediately arise with type-II observations. Are there any? And if there are, do they automatically support the theory they are laden with? Let me set the first question aside for a moment. The answer to the second question is 'No'. Mary Hesse (1980) gives a nice example to show this. It is an example which I already used in a different context. Recall that Anaximenes and Aristotle both believe objects fall down when unsupported, but Aristotle means 'to the centre of the earth', by 'down', while Anaximenes has some absolute sense of the term so that 'down' means in the same direction anywhere in the universe. Hesse asks us to imagine the two taken to Australia to test their theories. Upon release of an object, Aristotle reports his observation as an object falling down, which is just what his theory predicted; but Anaximenes says that he sees the object rise rather than fall down, the very opposite of what he expected. Thus, Anaximene's theory fails *in its own terms*. His observation is theory-laden, and is not tested by means of a (relatively) neutral perception, but it is *not* self-supporting.

The first question, 'Are there any type-II experiences?' is much harder to answer. I conjecture that there are. However, they are extremely hard to find; indeed, we may not be able to recognize them as theory-laden experiences until we have turned them into type-I observations by discovering something relatively neutral to check them against. The only examples I can think of are fictitious, like the Aristotle–Anaximenes example just mentioned, or the caloric example given on p.173.[7] Nevertheless, I will extend my conjecture and claim further, not only that there are type-II perceptions, but that

> *The theory-ladenness of a type-II perception is not recognized until it is turned into a type-I perception.*

This additional conjecture seems especially plausible if we consider examples such as the Müller-Lyer arrows. If we did not have an independent way of measuring the lengths of the two lines we would never have suspected that an illusion was present. We would think, quite reasonably, that one line is indeed longer than the other.[8]

NORMATIVE INTERVENTIONS

Are type-II observations and experiences present on a significant scale? That is, are there theory-laden observations for which there are (at present) no relatively neutral observations which could be used to evaluate them, and which, further, do not fail in their own terms as the Anaximenes example did? I don't want to take time here and now defending my conjecture that this is so; instead I want to outline a plan for coping with it, given that it is. This will be the crux of my anti-naturalist account of observation.

Let me begin the rest of my discussion by restating some major results of the last chapter. We cannot test theories by themselves against the data, as both the positivists and Popper thought. Rather, rational theory-appraisal is fundamentally and irreducibly comparative. We cannot objectively evaluate a single theory, only rank-order a collection of them. I take this here as given; it is surely one of the main results of recent philosophy of science. This comparative nature of appraisal requires a re-thinking of the discovery/justification distinction. The distinction was thought harmless in the past, since any social factors responsible for the invention of the theory in the first place could be filterd out in the process of justification, leaving an uncontaminated result. The justification process was seen as absolute, however, and we can no longer be so sanguine about the role of invention as we move to a comparative view of justification. As a card-carrying rationalist, I certainly believe that scientists can, and typically do, make rational choices. That is, they give an objective, rational ordering of any collection of available rival theories, and they accept the best from among these. These cognitive decisions are caused by good reasons, not by social forces.

There is a way of telling whether there might be unwitting social causation of our theories if we should find that the whole comparison group has some common property, a property which it is unlikely to have by mere chance. This is the sort of thing sociologists can ascertain. Of course, ascertaining it does not mean that we can conclude that 'social contamination' has occurred. We might, after all, discover that all the rival genetic theories were produced by left-handed scientists. Perhaps there is a connection, but it would seem a mere coincidence. On the other hand, most of the theories about women have been produced by men. This provides real grounds for suspicion. It is at this point, when we have made a sociological discovery, that a genuinely normative activity takes place. When we notice that there is a real possibility of having a skewed comparison group, it is then that we should make amends. We alter the character of the group itself. We can commission, so to speak, the production of more rivals, for example, rival theories about women produced this time by women. We can then return to the activity of rational theory-choice, which is still based on a comparison of those available, but the choice now, as far as we know, is no longer skewed. Of course, this sort of intervention, though it occurs sometimes, does not happen as often as I might like. I am speaking here more normatively than descriptively; this is how I think science can be made better. Hacking (1986) makes a similar point in his concerns with the impact on science of having so much research funded by a single source: defence.

At the outset I said that I was disenchanted with naturalism. If one looks at normal experimentation, observation, theory choice, and so on, one can probably do justice to a large bulk of it with a naturalistic account. However, part of the activity of good science is this fundamentally normative intervention in the process. This philosophical intervention sets it apart, I think, from anything that could justly be called 'naturalistic'. There is a sense, of course, in which everything that there is is natural, but that use of the term makes it vacuous. Giving a naturalistic account of the norms themselves won't help either. We normally distinguish between the flora in the wild and what we find in our cultivated gardens. The former is natural, the latter is not. There is no trouble in giving a naturalistic account of the growth of a tulip, but to leave it at that leaves too much untold. In the same way, I suggest, observation, experience, and all the rest of science, is much better understood as akin to a cultivated garden than to a pristine forest or a desert landscape.[9]

RECAPITULATION

It is now time to draw this book to a close with a brief summary. I began by sketching recent sociological approaches to science, especially the strong programme. Bloor's 'science of science' was criticized, as was Barnes's 'finitism'. Bloor held that, unlike rationalistic accounts of science, sociological approaches are 'scientific' and hence the right way to go. Barnes held that our 'finitistic' knowledge is conventional through and through. Neither of these views can hold up to careful scrutiny. Nor can such accounts of science as that produced by the 'anthropologists in the lab', Collins, and Latour and Woolgar. There is often much to be learned from the views of sociologists of science; their accounts are typically full of wonderful detail and genuine insight. But when morals are drawn about the structure and epistemology of science, about what really drives this glorious enterprise, sociological accounts fall flat on their faces.

This failure of sociological approaches leads us back to something like common sense – the common-sense opinion that science really is rational, not the common-sense opinion of what scientific rationality actually consists in. For finding out the latter, I propose a technique which utilizes the history of science. The idea is that we should try to make as many actions in the history of science as rational as possible, while at the same time doing justice to well-established sociological (anthropological, psychological) findings, which act as a kind of constraint.

Though the idea is simple in conception, its application to the history of science is no easy matter. A significant effort by historians, by philosophers, and (perhaps above all) by sociologists is required. It is, frankly, a large jump from outlining the technique to declaring a winner in the rationality sweepstakes. Nevertheless, I did sketch what I take scientific rationality to be. Perhaps the most important aspect of good scientific method is its comparative nature; rival theories are not evaluated outright as good or bad, but rather they are rank-ordered.

This in turn has lead to a reconsideration of the role of the social. Undeniably, the social can and does play a role in 'discovery'. If theory evaluation were absolute instead of comparative, then the social could be filtered out; but this is no longer a plausible view given a comparative account of theory appraisal (even if there is complete rationality and objectivity in the rank-ordering of rival theories).

To acknowledge a social component in science is to throw down the gauntlet. Once we know there could be a systematic bias in some theory (or class of theories), we are prodded into action. The present chapter has been an attempt to outline a kind of normative intervention into the scientific process; by this kind of intervention, science can be improved. Strong programmers like to see themselves as describing science without any sort of evaluation; they pose as the anthropologist who wants to study an exotic culture without having any influence upon it. But such a posture is really hopeless. If we see non-rational factors playing a significant role in the content of one of our beliefs, then that belief is undermined. Not only is it undermined in the normative epistemological sense, but it is undermined psychologically as well. If we can see how to make science better, then we shall and we will make it better. To paraphrase Marx once again, the sociologists have only interpreted science; the point is to improve it. To appreciate this is to go some way toward appreciating how the rational and the social fit together.

NOTES

CHAPTER ONE: THE SOCIOLOGICAL TURN

1 Philosophers typically use 'knowledge' to mean true justified belief. Accordingly, the sociology of knowledge might be better called 'the sociology of belief'. When it is of no particular significance, I may often follow the sociologists' custom of using the terms 'knowledge' and 'belief' interchangeably.
2 See, for example, Quine (1960), Kuhn (1970), Feyerabend (1975), Wittgenstein (1953).
3 Hesse (1980). See especially chapter two for her discussion of this issue.

CHAPTER TWO: THE SCIENCE OF SCIENCE

1 For a fuller discussion see Davis (1979), which includes a bibliography.
2 Alas, there is much more to be said on this issue. See, for example, Bloor (1984), Davidson (1980), Gutting (1984), Hollis (1982), Laudan (1981a), Newton-Smith (1982), and Taylor (1982).
3 I have added 'rational and irrational beliefs' to Bloor's formulation of the principle, because it is clear that he intends it, and it is important to do so for the sake of the following discussion.
4 Bloor, for instance, has convinced me that it does not apply to him.

CHAPTER THREE: FINITISM

1 See Reichenbach (1958), Grünbaum (1974), and Friedman (1983) for discussions.
2 The following argument is adapted from Nicholas (1984), where it is used very effectively.

CHAPTER FOUR: THE EXPERIMENTER'S SOCIAL CIRCLE

1 I follow Latour and Woolgar in using TRF(H), though TRH has become the standard term.
2 I am not assuming a hard and fast observational/theoretical distinction here, only one of degree, and only for the purposes of illustration.
3 Transversely Excited Atmospheric pressure CO_2 laser.
4 See Shapere (1982) for a good explication and defence of this.
5 Hacking himself is perfectly clear on the distinction.

CHAPTER FIVE: BOLINGBROKE VERSUS HENRY FORD

1 I should stress that as a very gifted historian himself, Garber takes the Henry Ford 'bunk' view of history only in the specific sense I am here concerned with: history as evidence for norms.
2 For further enlightening discussions on this issue see Butts (1980) and McAllister (1986).
3 For an interesting 'Hegelian' perspective on this see Hacking (1979).
4 For sympathetic treatments of Lakatos's view see Kourany (1982) and the excellent case studies in Howson (ed.) (1976).
5 In fact, this is exactly what Musgrave claims in his (1979) review of Laudan's *Progress and its Problems*.
6 For commentaries see Daniels (ed.) (1976) and especially Daniels (1979).
7 Grammarians propose general rules that try to capture speakers' intuitions about the grammatical correctness of particular sentences. Occasionally the general rule is used to reject the considered judgement on a specific example.
8 For this line of criticism see Hare (1976) or Singer (1974).

CHAPTER SIX: HOW TO BE AN ANTHROPOLOGIST OF SCIENCE

1 John Worrall (1976) has given an account which overlaps mine in many ways. See his excellent study for additional convincing reasons to adopt this general outlook.
2 This is a confusion which often besets a recent study by Broad and Wade (1982).
3 Martin Hollis (1982) says the principle of charity is a necessary 'bridgehead' for understanding different cultures. He is right: a bridgehead is needed; but it is based on our common rationality, not a common truth.
4 Before he adopted a naturalistic view.
5 I use 'presuppose' in the sense of material implication; 'p presupposes q' means 'If p then q'.

CHAPTER SEVEN: MAKING SCIENCE BETTER

1 I often hear in conversation: 'Science is arrogant and technology destructive. Three cheers for the strong programme since it brings science down a notch or two and so slows the implementation of new technology.'

2 Strictly, it is a definition of ideological determination, as he puts it, so ideology can then be defined as a belief that is ideologically determined.

3 For more, see: Bleier (1984, 1986), Harding (1986), Harding and Hintikka (1983), Keller (1983, 1985), Fausto-Sterling (1985).

4 The fight now is usually between realists who say the boundary is not ontologically significant and the instrumentalists who say it is. But that is another story.

5 Chapter two of this book contains a very good discussion of the theory-ladenness of perception. The example I draw from is on pp. 16ff.

6 Experiments have other uses than just testing theories, but their employment in theory evaluation is the use I am interested in here.

7 Some of the more realistic examples given by Kuhn and Feyerabend are, of course, possible candidates.

8 I am not suggesting, however, that all theory-laden experiences are just simple illusions.

9 The analogy should not be misunderstood as suggesting that the objects of science are artefacts. Gardens and science are both artefacts, but the *contents* of a garden are *natural* objects, and the same can be said of the objects of science. There is no conflict with scientific realism.

BIBLIOGRAPHY

Agassi, J. (1963) 'Towards an historiography of science', *History and Theory* 2:1–23.

Ayim, M. (1982) 'Wet sponges and bandaids – a gender analysis of speech patterns', *Semiotics Society of America*.

Barber, B. (1952) *Science and the Social Order*, New York: Free Press.

Barber, B. (1961) 'Resistance by scientists to scientific discovery', *Science* 134(3479):596–602.

Barber, B. and Hirsch, W. (eds) (1962) *The Sociology of Science*, New York: Free Press.

Barnes, S. B. (ed.) (1972) *Sociology of Science*, London: Penguin.

Barnes, S. B. (1974) *Scientific Knowledge and Sociological Theory*, London: Routledge & Kegan Paul.

Barnes, S. B. (1976) 'Natural rationality: a neglected concept in the social sciences', *Philosophy of the Social Sciences* 6(2):115–26.

Barnes, S. B. (1977) *Interests and the Growth of Knowledge*, London: Routledge & Kegan Paul.

Barnes, S. B. (1979) 'Vicissitudes of belief' (Review of Laudan 1977), *Social Studies of Science* 9:247–63.

Barnes, S. B. (1980) 'On the causal explanation of scientific judgement', *Social Science Information* 19:685–95.

Barnes, S. B. (1981) 'On the conventional character of knowledge and cognition', *Philosophy of the Social Sciences* 11(3):303–33.

Barnes, S. B. (1981a) 'On the 'hows' and 'whys' of cultural change', *Social Studies of Science* 11(4):481–98.

Barnes, S. B. (1982) *T. S. Kuhn and Social Science*, New York: Columbia University Press.

Barnes, S. B. (1982a) 'On the extensions of concepts and the growth of knowledge', *Sociological Review* 30(1):23–44.

Barnes, S. B. (1985) *About Science*, Oxford: Blackwell.

Barnes, S. B. and Bloor, D. (1982) 'Relativism, rationalism, and the sociology of knowledge', in M. Hollis and S. Lukes (1982).

Barnes, S. B. and Dolby, R. G. A. (1970) 'The scientific ethos: a deviant viewpoint', *European Journal of Sociology* 11(1):3–25.

Barnes, S. B. and Edge, D. (eds) (1982) *Science in Context: Readings in the Sociology of Science*, Boston: MIT Press.

Barnes, S. B. and MacKenzie, D. (1979) 'On the role of interests in scientific change', *Sociological Review Monographs* (Special Issue on 'Rejected Knowledge').

Barnes, S. B. and MacKenzie, D. (1979a) 'Scientific judgment in the Biometry–Mendelism controversy', in Barnes and Shapin (1979).

Barnes, S. B. and Shapin, S. (1977) 'Science, nature and control: interpreting mechanics' institutes', *Social Studies of Science* 7(1):31–74.

Barnes, S. B. and Shapin, S. (eds) (1979) *Natural Order*, London: Sage.

Barnes, S. B. and Shapin, S. (1979a) 'Darwin and social Darwinism: purity and history', in Barnes and Shapin (1979).

Ben-David, J. (1971) *The Scientist's Role in Society*, Englewood Cliffs, N.J.: Prentice-Hall.

Berger, P. and Luckmann, T. (1967) *The Social Construction of Reality*, London: Allen Lane.

Birke, L. (1986) *Women, Feminism and Biology*, Brighton: Harvester Press.

Bleier, R. (1984) *Science and Gender: a Critique of Biology and its Theories on Women*, New York: Pergamon.

Bleier, R. (1986) *Feminist Approaches to Science*, New York: Pergamon.

Bloor, D. (1971a) Two paradigms for scientific knowledge?', *Science Studies* 1:101–115.

Bloor, D. (1973) 'Wittgenstein and Mannheim on the sociology of mathematics', *Studies in History and Philosophy of Science* 4(2):173–91.

Bloor, D. (1973a) 'Are philosophers averse to science?', in D. Edge and J. Wolfe (eds) *Meaning and Control*, London: Tavistock, 1–17.

Bloor, D. (1974) 'Popper's mystification of objective knowledge', *Science Studies* 4:65–76.

Bloor, D. (1975) 'Psychology or epistemology', *Studies in History and Philosophy of Science* 6:382–95.

Bloor, D. (1976) *Knowledge and Social Imagery*, London: Routledge & Kegan Paul.

Bloor, D. (1978) 'Polyhedra and the abominations of Leviticus', *British Journal for the History of Science* 11:245–72.

Bloor, D. (1981) 'The strengths of the strong programme', *The Philosophy of the Social Sciences* 11:199–213 [Reprinted in J. R. Brown (1984)].

Bloor, D. (1981a) 'Hamilton and Peacock on the essence of algebra', in H. Mehrtens, H. Bos, and I. Schneider (eds) *Social History of the Nineteenth-Century Mathematics*, Boston, Basel and Stuttgart: Birkhauser, 202–32.

Bloor, D. (1982) 'Durkheim and Mauss revisited: classification and the sociology of knowledge', in *Studies in History and Philosophy of Science* 13(4):267–97.

Bloor, D. (1983) *Wittgenstein: a Social Theory of Knowledge*, London: Macmillan.

Bloor, D. (1984) 'The sociology of reasons: or why "epistemic factors" are really social factors,' in J. R. Brown (1984).

Bloor, D. and Barnes, B. (1982) 'Rationalism, relativism and the sociology of knowledge', in M. Hollis and S. Lukes (1982).

Bloor, D. and Bloor, C. (1982) 'Twenty industrial scientists: a preliminary exercise', in M. Douglas (ed.) *Essays in the Sociology of Perception*, London: Routledge & Kegan Paul, 83–102.

Boyd, R. (1973) 'Realism, underdetermination, and a causal theory of knowledge', *Nous*.

Brannigan, A. (1981) *The Social Basis of Scientific Discoveries*, Cambridge: Cambridge University Press.

Brant, R. (1959) *Ethical Theory*, Englewood Cliffs, N.J.: Prentice-Hall.

Broad, W. and Wade, N. (1982) *Betrayers of the Truth*, New York: Simon & Schuster.

Brown, J. R. (ed.) (1984) *Scientific Rationality: the Sociological Turn*, Dordrecht: Reidel.

Brown, T. M. (1974) 'From mechanism to vitalism in eighteenth-century English physiology', *Journal of the History of Biology* 7:179–216.

Butts, R. E. (1980) 'Methodology and the Functional Identity of Science and Philosophy', in J. Hintikka, D. Gruender and J. Agazzi (eds) *Pisa Conference Proceedings* II:253–70.

Campbell, D. (1973) 'Evolutionary epistemology', in P. A. Schillp (ed.) *The Philosophy of Karl Popper*, La Salle, Illinois: Open Court.

Cantor, G. N. (1975) 'Phrenology in early nineteenth-century Edinburgh: an historiographical discussion', *Annals of Science* 32:195–218.

Cantor, G. N. (1975a) 'A critique of Shapin's social interpretation of the Edinburgh phrenology debate', *Annals of Science* 33:245–56.

Cesi and Peters [a report] (1980) *Science* (5 September 1980).

Churchland, P. (1979) *Scientific Realism and the Plasticity of Mind*, Cambridge: Cambridge University Press.

Cole, J. R. and Cole, S. (1973) *Social Stratification in Science*, Chicago and London: University of Chicago Press.

Collins, H. M. (1974) 'The TEA set: tacit knowledge and scientific networks', *Science Studies* 4:165–85.

Collins, H. M. (1975) 'The seven sexes: a study in the sociology of a phenomenon, or the replication of experiments in physics', *Sociology* 9:205–24.

Collins, H. M. (1981) 'The social destruction of gravitational radiation', *Social Studies of Science* 11(1):33–62.

Collins, H. M. (1985) *Changing Order: Replication and Induction in Scientific Practice*, London: Sage.

Collins, H. M. and Pinch, T. J. (1979) 'The construction of the paranormal: nothing unscientific is happening', in R. Wallis (ed.) *On the Margins of Science*, 237–70, Keele, Staffs: Sociological Review Monograph 27.

Collins, H. M. and Pinch, T. J. (1982) *Frames of Meaning*, London: Routledge & Kegan Paul.

Coleman, W. (1970) 'Bateson and chromosomes: conservative thought in science', *Centaurus* 15(3–4):228–314.

Covell, K. and Brown, J. R. (1987) 'The nature and rationality of Piaget's revolution', *Methodology and Science*.

Crane, D. (1972) *Invisible Colleges*, Chicago: University of Chicago Press.

Cunningham, F. (1973) *Objectivity in Social Science*, Toronto: University of Toronto Press.

Daniels, N. (ed.) (1976) *Reading Rawls*, New York: Basic Books.

Daniels, N. (1979) 'Wide reflective equilibrium and theory acceptance in ethics', *Journal of Philosophy* LXXVI (May).

Davidson, D. (1973) 'The very idea of a conceptual scheme', *Proceedings of the American Philosophical Association* 1973–74:18ff.

Davidson, D. (1980) *Essays on Actions and Events*, Oxford: Oxford University Press.

Davis, L. (1979) *Theory of Action*, Englewood Cliffs: Prentice-Hall.

Douglas, Mary (1966) *Purity and Danger: an Analysis of Concepts of Pollution and Taboo*, London: Routledge & Kegan Paul.

Douglas, Mary (1970) *Natural Symbols*, London: Barrie and Jenkins.

Duhem, P. (1962) *The Aim and Structure of Physical Theory*, New York: Atheneum.

Durkheim, E. and Mauss, M. (1903) *Primitive Classification*, Eng. trans. (1963), London: Cohen & West.

Easlea, B. (1973) *Liberation and the Aims of Science*, London: Chatto & Windus.

Edge, D. O. and Mulkay, M. J. (1976) *Astronomy Transformed*, New York: Wiley Interscience.

Eichler, M. and Lapointe, J. (1985) *On the Treatment of the Sexes in Research*, Ottawa: Social Sciences and Humanities Resaerch Council of Canada.

Farley, J. (1977) *The Spontaneous Generation Controversy from Descartes to Oparin*, Baltimore: Johns Hopkins University Press.

Farley, J. and Geison, C. (1974) 'Science, politics and spontaneous generation in nineteenth-century France: the Pasteur–Pouchet debate', *Bulletin of the History of Medicine* 48:161–98.

Fausto-Sterling, A. (1985) *Myths of Gender*, New York: Basic Books.

Feuer, L. S. (1974) *Einstein and the Generations of Science*, New York: Basic Books.

Feyerabend, P. (1975) *Against Method*, London: New Left Books.

Fleck, L. (1935) *Entstehung und Entwicklung einer Wissenschaftliche Tatsache*, English translation by F. Bradley and T. J. Trenn (1979) *Genesis and Development of a Scientific Fact*, Chicago: University of Chicago Press.

Fodor, J. (1984) 'Observation re-considered', *Philosophy of Science*.

Forman, P. (1971) 'Weimar culture, causality and quantum theory, 1918–1927: adaptation by German physicists and mathematicians to a hostile intellectual environment', in R. McCormmach (ed.) *Historical Studies in the Physical Sciences* 3, Philadelphia, Penn.: University of Pennsylvania Press.

Frankel, E. (1976) 'Corpuscular optics and the wave theory of light: the science and politics of a revolution in physics', *Social Studies of Science* 6:141–84.

Frege, G. (1892) 'On sense and reference', in P. Geach and M. Black (eds)

(1952) *Translations from the Philosophical Writings of Gottlob Frege*, Oxford: Blackwell.

Friedman, M. (1983) *Foundations of Space–time Theories*, Princeton: Princeton University Press.

Garber, D. (1986) 'Learning from the past', *Synthese*, 91–114.

Gaston, J. (1973) *Originality and Competition in Science*, Chicago and London: University of Chicago Press.

Gaston, J. (1978) *The Reward System in British and American Science*, New York: Wiley.

Gaston, J. (ed.) (1978a) *The Sociology of Science*, San Francisco: Jossey-Bass.

Giere, R. (1973) 'History and philosophy of science: an intimate connection or a marriage of convenience?', *British Journal for the Philosophy of Science*.

Giere, R. (1984) 'Towards a unified theory of science', in J. Cushing, N. Delaney, and G. Gutting (eds) *Science and Reality*, Notre Dame: Notre Dame University Press.

Graham, L. (1985) 'The socio-political roots of Boris Hessen: Soviet marxism and the history of science', *Social Studies of Science*.

Grandy, R. (1973) 'Reference, meaning, and belief', *Journal of Philosophy*.

Gregory, R. (1970) *The Intelligent Eye*, New York: McGraw Hill.

Grünbaum, A. (1974) *Philosophical Problems of Space and Time*, second edition, Dordrecht: Reidel.

Gutting, G. (1984) 'The strong programme: a dialogue', in J. R. Brown (1984).

Habermas, J. (1972) *Knowledge and Human Interests*, London: Heinemann.

Hacking, I. (1979) 'Imré Lakatos's philosophy of science', *British Journal for the Philosphy of Science*, 381–410.

Hacking, I. (ed.) (1981) *Scientific Revolutions*, Oxford: Oxford University Press.

Hacking, I. (1983) *Representing and Intervening*, Cambridge: Cambridge University Press.

Hacking, I. (1984) 'Experimentation and scientific realism', in Leplin (ed.) (1984).

Hacking, I. (1986) 'Weapons research and the form of scientific knowledge', in D. Copp (ed.) *Nuclear Weapons, Deterrence, and Disarmament*, supplementary volume, *Canadian Journal of Philosophy*.

Hall, R. (1971) 'Can we use the history of science to decide between competing methodologies?', in C. Buck and R. Cohen (eds) *PSA 1970*, Dordrecht: Reidel.

Hanson, N. R. (1958) *Patterns of Discovery*, Cambridge: Cambridge University Press.

Hanson, N. R. (1961) 'Is there a logic of scientific discovery?', in H. Feigl, and G. Maxwell (eds) *Current Issues in the Philosophy of Science*, New York: Holt, Rinehart & Winston.

Hare, R. (1976) 'Rawls' theory of justice' in Daniels (1976).

Harding, S. (1986) *The Science Question in Feminism*, Ithaca: Cornell.

Harding, S. and Hintikka, M. (1983) *Discovering Reality*, Dordrecht: Reidel.

Hempel, C. G. (1965) *Aspects of Scientific Explanation*, New York and

London: The Free Press.

Hendry, J. (1980) 'Weimar culture and causality', *History of Science* xviii.

Hesse, M. B. (1974) *The Structure of Scientific Inference*, London: Macmillan.

Hesse, M. B. (1980) *Revolutions and Reconstructions in the Philosophy of Science*, Brighton: Harvester Press.

Hessen, B. (1931) 'The social and economic roots of Newton's *Principia*', in N. I. Bukharin *et al.* (eds) *Science at the Crossroads*, 2nd edition (1971), London: Frank Cass.

Hollis, M. (1982) 'The social destruction of reality', in Hollis and Lukes (1982).

Hollis, M. and Lukes, S. (eds) (1982) *Rationality and Relativism*, Oxford: Blackwell.

Holton, G. (1974) 'On being caught between Apollo and Dionysus', *Deadalus* (summer).

Hooker, C. (1986) *A Realistic Theory of Science*, Buffalo: Sunny.

Howson, C. (ed.) (1976) *Method and Appraisal in the Physical Sciences*, Cambridge: Cambridge University Press.

Hubbard, R. (1982) 'Have only men evolved?' in R. Hubbard, M. Henifin, and B. Fried (eds) *Biological Woman – The Convenient Myth*, Cambridge, Mass.: Schenman.

Jacob, J. R. (1977) *Robert Boyle and the English Revolution*, Brighton: Harvester Press.

Keller, E. F. (1983) *A Feeling for the Organism*, Chicago: Freeman.

Keller, E. F. (1985) *Reflections on Gender and Science*, New Haven: Yale University Press.

Klotz, I. M. (1980) 'The N-ray affair', *Scientific American*.

Kourany, J. (1982) 'Towards an empirically adequate theory of science', *Philosophy of Science*.

Kraft, P. and Kroes, P. (1984) 'Adaptation of scientific knowledge to an intellectual environment', *Centarus* 27:76–99.

Kripke, S. (1980) *Naming and Necessity*, Oxford: Blackwell.

Kuhn, T. S. (1957) *The Copernican Revolution*, Cambridge, Mass.: Harvard University Press.

Kuhn, T. S. (1963) 'The function of dogma in scientific research', in A. C. Crombie (ed.) *Scientific Change*, London: Heinemann, 347–69.

Kuhn, T. S. (1970) *The Structure of Scientific Revolutions*, 2nd edn, Chicago: University of Chicago Press (first published in 1962).

Kuhn, T. S. (1971) 'Notes on Lakatos' in Buck and Cohen (eds) *PSA 1970*, Dordrecht: Reidel.

Kuhn, T. S. (1974) 'Second thoughts on paradigms', in F. Suppe (ed.) *The Structure of Scientific Theories*, Chicago: Illinois University Press, 459–82 (reprinted in Kuhn 1977).

Kuhn, T. S. (1977) *The Essential Tension*, Chicago: University of Chicago Press.

Lakatos, I. (1970) 'Falsification and the methodology of scientific research programmes', in Lakatos and Musgrave (1970).

Lakatos, I. (1971) 'History of science and its rational reconstructions', in Buck and Cohen (eds) *PSA 1970*, Dordrecht: Reidel.

Lakatos, I. (1976) *Proofs and Refutations*, Cambridge: Cambridge University Press.

Lakatos, I. and Musgrave, A. (eds) (1970) *Criticism and the Growth of Knowledge*, Cambridge: Cambridge University Press.

Latour, B. and Woolgar, S. (1979) *Laboratory Life: the Social Construction of Scientific Facts*, London: Sage.

Laudan, L. (1977) *Progress and its Problems*, Berkeley: University of California Press.

Laudan, L. (1981) *Science and Hypothesis*, Dordrecht: Reidel.

Laudan, L. (1981a) 'The pseudo-science of science?' *Philosophy of the Social Sciences*, reprinted in J. R. Brown (1984).

Laudan, L. (1981b) 'A confutation of convergent realism', *Philosophy of Science*, 1981.

Laudan, L. (1984) *Science and Values*, Berkeley: University of California Press.

Laudan, L. (1986) 'Some problems facing intuitionist meta-methodologies', *Synthese*.

Laudan, L. (1987) 'Progress or rationality? The prospects for a normative naturalism', *American Philosophical Quarterly*.

Lelas, S (1985) 'Topology of internal and external factors in the development of knowledge', *Ratio* XXVII(1).

Leplin, J. (ed.) (1984) *Scientific Realism*, Berkeley: University of California Press.

Longino, H. and Doell, R. (1983) 'Body, bias, and behavior: a comparative analysis of responding in two areas of biological science', *Signs*.

Lukes, S. (1974) 'Relativism: cognitive and moral', *Proceedings of the Aristotelian Society*, supplementary vol. 48:165–89.

Lukes, S. (1982) 'Relativism in its place', in Hollis and Lukes (1982).

McAllister, J. (1986) 'Theory-assessment in the historiography of science', *British Journal for the Philosophy of Science*.

MacKenzie, D. (1978) 'Statistical theory and social interests: a case study', *Social Studies of Science* 8:35–83.

MacKenzie, D. (1979) 'Karl Pearson and the professional middle class', *Annals of Science* 36:125–43.

MacKenzie, D. (1981) *Statistics in Britain, 1865–1930: The Social Construction of Scientific Knowledge*, Edinburgh: Edinburgh University Press.

MacKenzie, D. and Barnes, S. B. (1975) 'Biometrician versus Mendelian: a controversy and its explanation', *Kölner Zeitschrift für Soziologie*, Special Edition no. 18:165–96.

MacKenzie, D. and Barnes, S. B. (1979) 'Scientific judgement: the Biometry–Mendelism controversy', in Barnes and Shapin (1979).

McMullin, E. (1970) 'History of science: a taxonomy' in *Minnesota Studies in the Philosophy of Science* IV.

McMullin, E. (1984) 'The rational and the social in the history of science' in J. R. Brown, (1984).

Mamchur, E. (1985) 'The principle of "maximum inheritance" and the growth of knowledge', *Ratio* XXVII(1).

Mannheim, K. (1936) *Ideology and Utopia*, New York: Harcourt, Brace & World.

Mannheim, K. (1952) *Essays on the Sociology of Knowledge*, London: Routledge & Kegan Paul.

Marx, K. (1859) *A Contribution to the Critique of Political Economy*, as translated from the German (1970), New York: International Publishers.

Maull, N. (1976) 'Reconstructed history as philosophical evidence', in F. Suppe and P. Asquith (eds) *PSA 1976*, Dordrecht: Reidel.

Maxwell G. (1962) 'The ontological status of theoretical entities; *Minnesota Studies in the Philosophy of Science* III.

Mellor, D. H. (1977) 'Natural kinds', *British Journal for the Philosphy of Science*.

Merton, R. K. (1957) *Social Theory and Social Structure* (revised and enlarged ed.), New York: The Free Press.

Merton, R. K. (1970) *Science, Technology and Society in Seventeenth-Century England*, New York: Harper & Row (originally published in 1938).

Merton, R. K. (1973) *The Sociology of Science*, Chicago and London: University of Chicago Press (ed. with an introduction by N. W. Storer).

Misner, C., Thorne, K. and Wheeler, J. (1973) *Gravitation*, Chicago: Freeman.

Mitroff, I. I. (1974) *The Subjective Side of Science*, Amsterdam: Elsevier.

Mulkay, M. J. (1974) 'Methodology in the sociology of science', *Social Science Information* 13:107–19.

Mulkay, M. J. (1976) 'Norms and ideology in science', *Social Science Information* 15:637–56.

Mulkay, M. J. (1976a) 'The mediating role of the scientific elite', *Social Studies of Science*, 6:445–70.

Mulkay, M. J. (1977) 'Connections between the quantitative history of science, the social history of science and the sociology of science, *Proceedings of the International Seminar on Science Studies*, Helsinki: Academy of Finland, 54–76.

Mulkay, M. J. (1977a) 'Sociology of the scientific research community', in I. Spiegel-Rosing and D. J. de Solla Price (eds) *Science, Technology and Society*, London: Sage, 93–148.

Mulkay, M. J. (1979) *Science and the Sociology of Knowledge*, London: Allen & Unwin.

Mulkay, M. J., Gilbert, G. N., and Woolgar, S. (1975) 'Problem areas and research networks in science', *Sociology* 9:187–203.

Mulkay, M. J. and Williams, A. T. (1971) 'A sociological study of a physics department', *British Journal of Sociology* 22:68–82.

Musgrave, A. (1979) 'A problem with progress', *Synthese* 42:443–64.

Nagel, E. (1961) *The Structure of Science*, London: Routledge & Kegan Paul.

Newton-Smith, W. (1981) *The Rationality of Science*, Oxford: Routledge & Kegan Paul.

Newton-Smith, W. (1982) 'Relativism and the possibility of interpretation', in Hollis and Lukes (1982).

Nicholas, J. (1984) 'Scientific and other interets', in J. R. Brown (1984).

Nickles, T. (ed.) (1980) *Scientific Discovery* (2 vols.), Dordrecht: Reidel.

Okruhlik, K. (forthcoming, a) 'The locus of value in science'.

Okruhlik, K. (forthcoming, b) Review of Harding (1986), *Philosophy of Science*.

Pickering, A. (1981) 'The role of interests in high-energy physics: the choice between charm and colour', *Sociology of the Sciences Yearbook* 5:107–38.

Pickering, A. (1984) *Constructing Quarks*, Chicago: University of Chicago Press.

Pinch, T. (1985) *Confronting Nature*, Dordrecht: Reidel.

Poincaré, H. (1952) *Science and Hypothesis*, New York: Dover.

Polanyi, M. (1958) *Personal Knowledge*, London: Routledge & Kegan Paul.

Popper, K. R. (1959) *The Logic of Scientific Discovery*, New York: Harper & Row.

Popper, K. R. (1960) *The Poverty of Historicism*, London: Routledge & Kegan Paul.

Popper, K. R. (1963) *Conjectures and Refutations*, London: Routledge & Kegan Paul.

Popper, K. (1972) *Objective Knowledge*, London: Oxford.

Price, D. J. de Solla (1963) *Big Science, Little Science*, New York: Columbia University Press.

Putnam, H. (1975) 'The meaning of meaning' in Putnam (1975–83).

Putnam, H. (1975–83) *Philosophical Papers* (3 vols.), Cambridge: Cambridge University Press.

Quine, W. (1935) 'Truth by convention', in *The Ways of Paradox*, New York: Random House.

Quine, W. (1953) 'Two dogmas of empiricism', in *From a Logical Point of View*, Cambridge, Mass.: Harvard University Press.

Quine, W. (1960) *Word and Object*, Boston: Harvard University Press.

Quine, W. (1969a) 'Epistemology naturalized', in *Ontological Relativity and Other Essays*, New York: Columbia University Press.

Quine, W. (1969b) 'Natural kinds', in *Ontological Relativity and Other Essays*, New York: Columbia Univesity Press.

Rawls, J. (1971) *A Theory of Justice*, Cambridge, Mass.: Harvard University Press.

Ravetz, J. R. (1971) *Scientific Knowledge and its Social Problems*, Oxford: Clarendon Press.

Reichenbach, H. (1938) *Experience and Prediction*, Chicago: University of Chicago Press.

Reichenbach, H. (1958) *Philosophy of Space and Time*, New York: Dover.

Rescher, N. (1976) 'Peirce and the economy of research', *Philosophy of Science*.

Roll-Hansen, Nils (1979) 'Experimental method and spontaneous generation: the controversy between Pasteur and Pouchet, 1859–64, *Journal of the History of Medicine* 34:273–92.

Rose, H. and Rose, S. (eds) (1976) *The Political Economy of Science*, London: Macmillan.

Rudwick, M. J. S. (1972) *The Meaning of Fossils*, London: Macdonald.

Rudwick, M. J. S. (1974) 'Darwin and Glen Roy: a "great failure" in scientific method?', *Studies in the History of Philosophy of Science* 5:97–185.

Ruse, M. (1979) *The Darwinian Revolution*, Chicago: University of Chicago Press.

Ruse, M. (1981) *Is Science Sexist?* Dordrecht: Reidel.

Ruse, M. (1986) *Taking Darwin Seriously*, Oxford: Blackwell.

Russell, B. (1905) 'On denoting', *Mind*.

Salmon, W. (1970) 'Bayes's theorem and the history of science', *Minnesota Studies in the Philosophy of Science* V.

Sandow, A. (1938) 'Social factors in the origin of Darwinism', *The Quarterly Review of Biology* 13:315–26.

Scheffler, I. (1963) *The Anatomy of Inquiry*, New York: Knopf.

Scheffler, I. (1967) *Science and Subjectivity*, New York: Bobbs-Merrill.

Searle, J. (1970) *Speech Acts*, Cambridge: Cambridge University Press.

Shapere, D. (1982) 'The concept of observation in science and philosophy', *Philosophy of Science*.

Shapin, S. (1975) 'Phrenological knowledge and the social structure of early nineteenth-century Edinburgh', *Annals of Science* xxxii:219–43.

Shapin, S. (1979) 'Homo phrenologicus: anthropological perspectives on an historical problem', in Barnes and Shapin (1979).

Shapin, S. (1979a) 'The politics of observation: cerebral anatomy and social interests in the Edinburgh phrenology disputes', in Roy Wallis (ed.) *On the Margins of Science: the Social Construction of Rejected Knowledge*, Sociological Review Monographs xxvii (Keele, Staffs):139–78.

Shapin, S. (1980) 'Social uses of science', in G. S. Rousseau and Roy Porter (eds) *The Ferment of Knowledge: Studies in the Historiography of Eighteenth-Century Science*, Cambridge: Cambridge University Press, 93–139.

Shapin, S. (1981) 'Of gods and kings: natural philosophy and politics in the Leibniz–Clarke disputes', *Isis* lxxii:187–215.

Shapin, S. (1982) 'History of science and its sociological reconstructions', *History of Science* xx:157–211.

Shapin, S. and Barnes, S. B. (1977) 'Science, nature and control: interpreting mechanics' institutes', *Social Studies of Science* 7(1):31–74.

Shapin, S. and Barnes, S. B. (1979) 'Darwin and social Darwinism: purity and history', in Barnes and Shapin (1979).

Shapin, S. and Schaffer, S. (1985) *Leviathan and the Air Pumps: Hobbes, Boyle and the Politics of Experiment*, Princeton: Princeton University Press.

Singer, P. (1974) 'Sidgwick and reflective equilibrium', *Monist*, 490–517.

Sklair, L. (1973) *Organized Knowledge*, London: Hart-Davis, MacGibbon.

Stark, W. (1958) *The Sociology of Knowledge*, London: Routledge & Kegan Paul.

Stark, W. (1967) 'The sociology of knowledge' in Edwards (ed.) *Encyclopedia of Philosophy*, New York: Macmillan.

Storer, N. W. (1966) *The Social System of Science*, New York: Holt, Rinehart & Winston.

Stroud, B. (1981) 'The significance of naturalized epistemology', *Midwest Studies in Philosophy* VI.

Suppe, F. (1977) 'Introduction and Afterward', in Suppe (ed.) *The Structure of Scientific Theories*, second edition, Chicago: University of Illinois Press.

Taylor, C. (1982) 'Rationality', in Hollis and Lukes (1982).

Teich, M. and Young, R. M. (eds) (1973) *Changing Perspectives in the History of Science*, London: Heinemann.

Thackray, A. (1974) 'Natural knowledge in cultural context', *American History Review* 74:672–709.

Tobey, R. C. (1971) *The American Ideology of National Science*, Pittsburgh, Pa.: University of Pittsburgh Press.

Toulmin, S. (1972) *Human Understanding*, Princeton: Princeton University Press.

Turner, R. S. (1971) 'The growth of professorial research in Prussia, 1818 to 1848 – causes and context', in R. McCormmach (ed.) *Historical Studies in the Physical Sciences* 3, Philadelphia, Penn.: University of Pennsylvania Press.

van Fraassen, B. (1980) *The Scientific Image*, Oxford: Oxford University press.

Watson, J. (1968) *The Double Helix*, New York: Atheneum.

Weaver, W. (1977) *Lady Luck*, New York: Dover.

Whewell, W. (1840) *Philosophy of the Inductive Sciences*, London.

Wilson, N. (1959) 'Substances without substrata', *Review of Metaphysics*.

Winch, P. (1958) *The Idea of a Social Science*, London: Routledge & Kegan Paul.

Winch, P. (1964) 'Understanding a primitive society', *American Philosophical Quarterly* 1:307–24.

Winsor, M. P. (1976) *Starfish, Jellyfish, and the Order of Life*, New Haven: Yale University Press.

Wisdom, J. O. (1953) *The Unconscious Origins of Berkeley's Philosophy*, London: Hogarth Press.

Wittgenstein, L. (1953) *Philosphical Investigations*, Oxford: Blackwell.

Wittgenstein, L. (1964) *Remarks on the Foundations of Mathematics*, Oxford: Blackwell.

Woolgar, S. W. (1976) 'Writing on intellectual history of scientific development: the use of discovery accounts', *Social Studies of Science* 6:395–422.

Woolgar, S. W. (1976a) 'The identification and definition of scientific collectivities', in G. Lemaine *et al.* (eds) *Perspectives on the Emergence of Scientific Disciplines*, Paris: Mouton, and Chicago: Aldine.

Woolgar, W. and Latour, B. (1979) *Laboratory Life: The Social Construction of Scientific Facts*, London: Sage.

Worrall, J. (1976) 'Thomas Young and the "Refutation" of Newtonian optics' in Howson (1976).

Wilson, B. (ed.) (1971) *Rationality*, Oxford: Blackwell.

Wylie, A. (1987) Review of Harding (1986) in M. Hanen and K. Nielson (eds) *Science, Morality, and Feminist Theory* supplementary volume, *Canadian Journal of Philosophy*.

Yearley, S. (1982) 'The relationship between epistemological and sociological cognitive interests: some ambiguities underlying the use of interest theory in the study of scientific knowledge', *Studies in the History and Philosophy of Science*.

Young, R. M. (1969) 'Malthus and the evolutionists', *Past and Present* 43:109–45.

Young, R. M. (1973) 'The historiographic and ideological contexts of the nineteenth-century debate on man's place in nature', in Teich and Young (eds) (1973).

193

INDEX